VOLCANOLOGY

VOLCANOLOGY

SECOND EDITION

Jacques-Marie Bardintzeff
Université de Paris, Orsay, France

and

Alexander McBirney
University of Oregon, Eugene, Oregon

WITH DRAWINGS BY CHRISTINE RYAN

JONES AND BARTLETT PUBLISHERS
Sudbury, Massachusetts
BOSTON   TORONTO   LONDON   SINGAPORE

World Headquarters
Jones and Bartlett Publishers
40 Tall Pine Drive
Sudbury, MA 01776
978.443.5000
info@.jbpub.com
www.jbpub.com

Jones and Bartlett Publishers Canada
2406 Nikanna Road
Mississauga, Ontario
Canada L5C 2WG

Jones and Bartlett Publishers International
Barb House, Barb Mews
London W6 7PA
UK

Library of Congress Cataloging-in-Publication Data

Bardintzeff, Jacques-Marie.
   [Volcans. English]
   Volcanology/Jacques-Marie Bardintzeff and Alexander McBirney; with drawings by
      Christine Ryan.
   p. cm.
   Includes bibliographical references and index.
   ISBN 0-7637-1318-X
   1. Volcanoes. 2. Volcanism. I. McBirney, Alexander R. II. Title
   QE522 .B3813 2000
   551.21—dc21
                            99-088172

Production Credits
Chief Executive Officer: Clayton Jones
Chief Operating Officer: Don Jones, Jr.
V.P., Sales and Marketing: Tom Manning
V.P., College Editorial Director: Brian L. McKean
V.P., Managing Editor: Judith H. Hauck
V.P., Design and Production: Anne Spencer
Senior Marketing Manager: Jennifer M. Jacobson
Production Editor: Rebecca S. Marks
Editorial/Production Assistant: Tim Gleeson
Director of Manufacturing and Inventory Control: Therese Bräuer
Cover Design: Stephanie Torta
Design and Composition: Graphic World Inc.
Printing and Binding: Malloy
Cover Printing: Malloy

This work has been published with the help of the French Ministère de la Culture–Centre national du livre.

About the cover: A strombolian eruption of Piton Kapor on Reunion Island on the night of the 16th of April 1998. Incandescent bombs were ejected to heights of 50 meters. Photo by JM Bardintzeff.

Printed in the United States of America
04 03 02 01 00   10 9 8 7 6 5 4 3 2 1

FOREWORD

▲

Originally published in two French editions under the name of the first author, Jacques-Marie Bardintzeff (*Volcanologie*, Masson, Paris, 1991; Dunod, Paris, 1998), the present English edition has been extensively updated and revised for an American audience. Geological examples in North America and Hawaii have been substituted for examples from Europe and the French overseas territories that are less familiar to American readers. Many of the new illustrations were taken from the more advanced-level text of Williams and McBirney, *Volcanology*, published by Freeman in 1979 and available from the publishers of the present edition, Jones and Bartlett.

We are indebted to Cathy Cashman (University of Oregon), Wendell Duffield (U.S. Geological Survey), Julius Dasch (NASA), and Derek Bostok for their helpful reviews of the preliminary manuscript and to the many individuals who have shared their knowledge and experience with us. While we have tried to credit the sources of our illustrations, we have not been able to do this for certain photographs or diagrams obtained many years ago.

This work has been published with the help of the French Ministère de la Culture—Centre National du Livre.

Although this text is designed for the undergraduate student in Geology, it should be comprehensible to any educated person with an interest in the natural sciences.

Jacques-Marie Bardintzeff
Alexander McBirney

ABOUT THE
AUTHORS

▲

Jacques-Marie Bardintzeff (right) has been teaching at the Orsay branch of the University of Paris since he obtained his doctorate at the same institution in 1985. He has studied the eruptive styles and hazards of volcanoes in the West Indies, Central America, Indonesia, and many other parts of the world. He is the author of several books published in France.

Alexander McBirney (left) is an emeritus professor at the University of Oregon. During his long career, he has worked in Central America, the Cascades, and the Galapagos Islands and for many years has carried out petrologic studies of the Skaergaard Intrusion of East Greenland. He has written several books on volcanoes and igneous rocks.

CONTENTS

▲

INTRODUCTION

▲

The Challenge of Volcanology

How can one be indifferent to volcanoes? Their beauty and awe-inspiring menace have always been a source of fascination and wonder. The earliest humans living in the Rift Valley of East Africa must have been just as impressed by eruptions of Mt. Kenya and Kilimanjaro as our European ancestors were by Etna or Vesuvius, or as we are today by Kilauea or Mt. St. Helens. Throughout history, people living in the shadow of volcanoes have given them a central place in legends and myths. For the Greeks and Romans, volcanoes were the abode of Héphaïstos (or Vulcan), the god of fire and metals. In Hawaii, they are the domain of the goddess Pelé, and for the Aztecs the god of fire, Huehueteotl. Even today, two million Japanese pilgrims, clad in white robes, climb the slopes on Fujiyama every year, and in Indonesia, offerings are made to Brahma in the crater of the volcano Bromo in hope that the faithful will be saved from an eruption during the coming year.

For centuries, volcanoes remained one of the least understood of natural phenomena. Most philosophers of the ancient world attributed volcanism to angry gods and "subterranean fires." Because they were held in such awe, volcanoes were rarely viewed objectively. The naturalist Pliny the Younger left a remarkably accurate account of the eruption of Vesuvius in AD 79 in which his uncle lost his life, but this was exceptional. Most accounts of volcanoes were based more on superstition and speculation than on accurate observations. It was not until the mid-18th century that naturalists began to appreciate the true nature of volcanism and what we now think of as modern volcanology emerged.

As so often happens in science, progress was the result of a lively debate. Abraham Gottlob Werner (1749–1817) proposed a widely accepted theory that all crystalline rocks, including basalt and granite, were precipitated from an ancient "Universal Ocean." Volcanoes, in his view, were rare anomalies resulting from subsurface combustion of coal. This idea went largely unquestioned until a handful of observant individuals working mainly in central France and Scotland recognized the high-temperature origin of igneous rocks and opened the way for modern volcanic geology. In 1763, Nicolas Desmarest traced a flow of columnar basalt to its origin in a volcanic crater near Clermont Ferrand and in so doing demonstrated that, contrary to a widely held belief, basalts are products of volcanic eruptions. Noting that some of the basalts had risen through granite, he realized that volcanoes could not be caused by burning coal beds, which, as Werner maintained, were found only at much shallower levels than granite.

James Hutton, an Edinburgh physician, came to similar conclusions. By showing that certain bodies of basalt and granite had intruded and thermally altered sedimentary beds, he undermined the accepted doctrine that granites were early precipitates from the sea and

Fig. 0.1 The ruins of the town of St. Pierre at the foot of Mt. Pelée after the eruption of 1902. (Photograph by A. Lacroix, 1904.)

were blanketed by younger clastic sediments. James Hall, a friend of Hutton and a professor of geology at the University of Edinburgh, succeeded in melting igneous rocks in a furnace and, after allowing them to cool slowly, showed that they solidified with the same crystalline form as natural rocks.

Subsequent progress in volcanology has been closely linked to major eruptions. In the chapters that follow, we shall note many basic concepts that were first recognized during certain unusual types of eruptions. For example, the catastrophic eruption of Mt. Pelée in 1902 led to the recognition of a particularly powerful type of explosive eruption in which turbulent flows of hot gas and fragmental material travel at great speed down the slopes of a volcano (Fig. 0.1). More recently, the 1980 eruption of Mt. St. Helens led to a new understanding of directed blasts and debris flows (Fig. 0.2). Several recent eruptions have heightened our awareness of the fundamental role of

Fig. 0.2 When Mt. St. Helens erupted in 1980, a hot pyroclastic surge swept down the valley of the Toutle River. It overtook the driver and passenger, who were traveling at full speed in an attempt to escape. (Photograph by A. R. McBirney.)

volcanism in shaping the earth and the environment in which we live. Most obvious, of course, is the heavy toll it takes on human life. At the same time, volcanism has been an important factor affecting the earth's climate and is responsible for a wealth of basic resources, including metals, construction materials, geothermal energy, and the fertility of soils.

Despite the impressive progress of recent years, we are still far from reaching an adequate understanding of volcanism. We can describe subtle compositional differences in magmas but can only speculate on their origin. We can measure temperatures, viscosities, and gas contents of magmas but cannot say why one volcano has an eruptive behavior dramatically different from that of a neighbor only a few kilometers away. We can determine the ages of ancient lavas but cannot say when or why a new eruption will occur, and in too many cases, we are unable to predict the nature of an eruption and the impact it is likely to have on the surrounding region. The challenge this poses can be illustrated by a brief review of a few recent eruptions.

The eruption of Mt. St. Helens, Washington, in the spring of 1980 was the first important volcanic event in North America in more than 60 years. Because it took place in a well-populated region where it could be closely monitored and studied, it stimulated a great surge of volcanological research. The eruption opened with mild steam eruptions that seemed to pose no danger. While the U. S. Geological Survey closely monitored the activity, the residents of a nearby resort area at Spirit Lake were urged to leave and authorities attempted to exclude the hundreds of journalists and casual observers who converged on the scene. During the following weeks, the northern slope began to swell as magma rose from depths and inflated the upper levels of the edifice. A bulge on the northern flank raised the surface as much as 150 m and eventually became unstable. At 8:32 on the morning of May 18th an earthquake of magnitude 5 triggered an avalanche of 2.3 km³ of rock, which released a powerful lateral explosion when the confining pressure on the underlying magma and hot water was removed. A powerful blast overwhelmed Spirit Lake and the surrounding hills, flattening the forest over an area of 600 km². A debris flow consisting of rock, mud, ice, and water swept down the mountain and continued down the Toutle River carrying huge quantities of sediment as far as the Columbia River. A 25-km high eruption column spread ash over much of eastern Washington and parts of Oregon, Idaho, Montana, and Wyoming. The magnitude of the blast and debris flow was totally unexpected. Despite the careful monitoring, scores of lives were lost, including that of David Johnson, a volcanologist who was observing the volcano from a ridge 13 km from the summit. The number of deaths would have been much greater if the region were not so sparsely settled and if access to the threatened area had not been limited.

The lessons learned at Mt. St. Helens helped volcanologists appreciate the potential dangers posed by the volcano Nevado del Ruiz in Columbia when it began to show signs of unrest in 1984. The snow-capped volcano, which had a long history of activity, was often referred to as *El Leon Dormido*—the sleeping lion. Unlike Mt. St. Helens, it was surrounded by a dense population with several towns and villages situated close to the flanks of the volcano. Geologists became concerned that melting of ice at the summit crater could result in devastating mudflows, and indeed this is exactly what took place. On the 13th of November 1985, a mild eruption of ash melted ice and snow near the summit and triggered a series of mudflows that swept down the flanks. Four major torrents of mud and debris were channeled into the steep valleys, carrying everything in their path. The town of Armero 60 km east of the summit was overwhelmed along with almost its entire population of 22,000 persons. The towns of Mariquita and Chinchina were struck by other flows descending the western slopes.

Much controversy followed the eruption. The danger posed by the volcano was clearly recognized; the town of Armero had been destroyed by a similar event in 1845. Although volcanologists had defined the hazards and prepared maps delineating a hazardous zone, no adequate warning system had been installed and communication proved tragically inadequate. On other volcanoes where hazards of this kind have been recognized, devices are installed on the upper slopes to detect mudflows automatically and send a signal to authorities so that organized evacuation can be carried out promptly. If such a system had been in place on Nevado del Ruiz, the 90 minutes that elapsed between the initiation of the mudflow and the time it struck the town of Armero would have been sufficient to evacuate the town and avoid the loss of so many lives.

As these examples suggest, the greatest toll of human lives has come, not from lava flows or ash falls, but from ash flows and lahars. (The latter are often referred to as mudflows, but they are more properly referred to as "lahars," the name used in Indonesia where they are particularly common.) And yet the variety of volcanic hazards is diverse and can seldom be anticipated. One of the most unusual examples was an eruption that took more than a thousand lives in the West African country of Cameroon in August of 1986. In this case, 1,746 victims died of asphyxiation.

Lake Nyos occupies a volcanic crater which, although it had no historical record of activity, was traditionally considered to possess evil powers. The water was saturated with carbon dioxide (CO_2) that over many years was emitted from vents on the lake bottom. Because the amount of CO_2 in solution increased with depth and pressure, it was very unstable. All that was required to trigger exsolution was a relaxation of pressure, possibly by an eruption or even by a change of atmospheric pressure and an overturn bringing saturated water to a shallower depth. This seems to be what happened in 1986. A sudden, short-lived eruption of gas threw the water to a height of several hundred meters and released a heavy cloud, a cubic kilometer in volume, consisting of 10 to 20% CO_2.

In this case, there was no way of anticipating the eruption; geologists had never seen an event of the kind before, and for many years, its cause was hotly debated. Studies are now being conducted to prevent further eruptions by artificially bringing the CO_2-rich bottom water to the surface, where it can exsolve under controlled conditions. It is hoped that monitoring of the lake can prevent future disasters.

Even when an effective warning system is in place and the threatened population is evacuated, it may be impossible to prepare for all possible eventualities. The Philippine volcano Pinatubo had been quiet for several centuries but was known to be dangerous. When a major eruption began in April of 1991, all reasonable precautions were taken to mitigate its effects. A quarter of a million persons were evacuated. Nevertheless, several hundred lives were lost, mainly to lahars. No one had anticipated the combination of events leading to the disaster.

During the early stages, at least 7 km^3 of ash were deposited on the flanks of the volcano. This in itself posed no danger, but starting on the 14th of June, the typhoon "Yunya" reached the island of Luzon, and its center passed only 50 km from the volcano on the 15th of June; others followed on the 18th and 19th. The torrential rains saturated the ash and set off voluminous lahars. The overwhelming scale of the phenomena made it impossible to evacuate the entire threatened region.

Volcanologists are understandably reluctant to recommend a wholesale evacuation of the populace around a threatening volcano. This is a drastic measure involving considerable cost and disruption of people's lives. Past experience has shown that if the population is evacuated and no eruption ensues, volcanologists are accused of being needlessly cautious. It is only natural that if activity continues sporadically for a prolonged period with no serious impact, the populace will conclude that the risk has been exaggerated. This was shown most dramatically during the recent eruptions of a volcano known as the Soufriere Hills on the small Caribbean island of Montserrat. Volcanologists were keenly aware of the danger posed by a dome that began to grow on the upper slopes. Hazardous areas were carefully delineated and plans for evacuation were thoroughly organized so that when an eruption seemed to be pending, the population was moved to safety. This happened three times and each time the crisis passed without an eruption. Volcanologists could make no firm predictions; they knew only that the volcano was in a state of unrest and, judging from experience with similar events in the past, recognized the danger of devastating ash flows coming from the flanks of the growing dome. Unfortunately, the repeated "false alarms" undermined the confidence of the public and made them increasingly reluctant to leave their homes when warnings were issued. As a result, when another explosive phase occurred on the 25th of June, 1997, about 20 persons who remained in the threatened zone needlessly lost their lives.

By far, the most serious challenge to volcanologists today is that of averting a major catastrophe when the Italian volcano Vesuvius erupts again. Although the volcano has not erupted since 1944, it has a long history of violent eruptions, such as the one that buried Pompeii and Herculaneum in AD 79 (Fig. 0.3). The magnitude of these eruptions tends to be greater after long periods of repose; the longer the volcano lies dormant, the more violent the next eruption is likely to be.

A population of more than a million persons now lives within the zone of potential impact of such an eruption. It is estimated that it would require a week to evacuate the region, but volcanologists doubt whether they can issue an alert on such a short time scale. They continue to study the volcano and search for clues that may enable them to provide a more precise warning and thereby avoid, or at least reduce, the horrible tragedy that threatens the metropolitan region of Naples.

Many other examples could be cited to illustrate the clear need for a better understanding of the origins and mechanisms of volcanism and the manner in which it can affect human lives. Although we have an extensive knowledge of the physics and chemistry of volcanoes, much remains to be learned, particularly in the realm of anticipating and mitigating their impact on humans. In the chapters that follow, we examine the origins and nature of volcanism by considering, first, why volcanoes are where they are and how molten magma generated in the earth's mantle finds its way to the surface. We then turn to the manifestations of volcanism in all its diverse forms—lavas, ash, and fragmental flows.

Fig. 0.3 The famous eruption of Vesuvius in AD 79 buried the city of Pompeii. At that time, only twenty thousand persons inhabited the region. Today, the metropolitan area of Naples extends over most of the area that would be affected if a similar eruption were to occur in the near future. There is no question that Vesuvius will erupt again. The question that volcanologists face is whether they can provide a warning in time to evacuate the population and avert a major catastrophe. (Photograph by J. M. Bardintzeff.)

Finally, we examine the impact of volcanism on the environment in which we live. While giving due attention to volcanic hazards, we examine some of the important benefits of volcanism, particularly the great energy resources that have only recently been exploited.

Suggested Reading

Fisher, R. V., G. Heiken, and J. B. Hulen. 1997. *Volcanoes, crucibles of change.* Princeton, NJ: Princeton University Press, 317 p.
  An excellent summary of the 1980 eruption of Mt. St. Helens (Chapter 1).

Lipman, P. W., and D. R. Mullineaux, eds. 1981. *The 1980 eruptions of Mount St. Helens. U S Geol Surv Prof Paper* 1250.
A collection of papers documenting all aspects of one of the most important eruptions of this century.

Sigurdsson, H., S. Carey, and R. S. J. Sparks. 1982. The eruption of Vesuvius in AD 79: Reconstruction from historical and volcanological evidence. *Am J Arch* 86:39–51.
An excellent example of how geological observations are used to reconstruct a major eruption.

Simkin, T., and R. S. Fiske. 1983. *Krakatau 1883: The volcanic eruption and its effects.* Washington, DC: Smithsonian Institution, 464 p.
A compilation of papers on all aspects of one the most famous eruptions of historic times.

Thomas, G., and M. M. Witts. 1969. *The day the world ended.* New York: Ballantine Books, 305 p.
A dramatic account of the eruption of Mt. Pelée and the destruction of St. Pierre and its 30,000 inhabitants. Although generally accurate, the authors perpetuate the myth that authorities discouraged the population from evacuating the city.

VOLCANOLOGY

▲

The Origins of Magmas

▼

1.1 ▲ Introduction

We are living in a period of unusually intense volcanism. A recently compiled catalog lists 1,511 volcanoes that are known to have erupted in the past 10,000 years, and about 30 of these erupt in any given year. This period of activity is by no means typical of the past. Although there have been times, such as the Miocene epoch, that were marked by even stronger volcanism, others, such as the Carboniferous periods, were notably more quiet. Certain times were dominated by plutonic intrusions with little volcanism; others the reverse. Although we have too little information to show this quantitatively, there is little doubt that volcanism is strongly episodic.

The spatial distribution of volcanoes is equally uneven. Most volcanoes are concentrated at the boundaries of tectonic plates, namely along oceanic spreading axes, continental rifts, and subduction zones. A few are found where continental plates collide without subduction, but these are rare. Almost all volcanoes that are not located near plate boundaries are the products of persistent thermal anomalies, or *hotspots*, in the interiors of both oceanic and continental plates. Hawaii and Yellowstone are perhaps the most familiar examples. To understand why volcanoes are distributed in this way, we must first examine their sources in the mantle and the processes by which magmas are generated.

1.2 ▲ The Mantle Origin of Magmas

It was long thought that the molten rock, or *magma*, that erupts as lava or ash came from a permanent reservoir of molten material in the

1

earth's interior. Although it is true that temperatures near the center of the earth approach 5000°C—a value far greater than that at which rocks melt at the surface (800 to 1200°C), it must be remembered that melting temperatures are greatly increased by the enormous pressure on the rocks at such depths. Records of seismic waves passing through the earth's interior show that by far the largest parts of the earth—the outermost crustal rocks, the underlying mantle, and the inner core— behave like solids (Fig. 1.1). Only the outer core between depths of 2,900 and 5,100 km is in a state approaching that of a liquid. We know this because transverse seismic waves are not transmitted through this zone. But this dense, iron-rich material making up the core is unlike any normal magma ever seen at the surface. The magmas erupted from volcanoes must be produced at some depth in the mantle or lower crust by melting of rocks that are normally solid.

The key factor in producing magmas is the earth's internal heat. Part of this heat is inherited from the original accretion of the planet, whereas another part is produced by decay of radioactive elements, chiefly potassium, thorium, and uranium, that are distributed in small amounts throughout the crust and mantle. Because rocks are such poor conductors, this steady addition of radiogenic heat cannot be dissipated by conduction to the surface as rapidly as it forms, and the temperature of the earth would be increasing with time were the man-

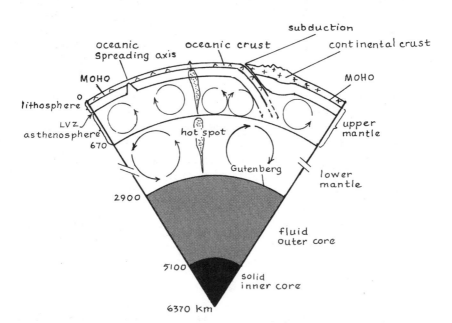

Fig. 1.1 A schematic section through the earth. For clarity, the thicknesses of shallow units have been exaggerated.

tle not able to turn over slowly by convection. Even though the rocks are nominally solid, at high temperatures and pressures, they are plastic enough to deform by viscous flow.

The temperature gradient of the earth's interior (or *geotherm*) is not linear (Fig. 1.2). Temperatures in the crust increase with depth at a

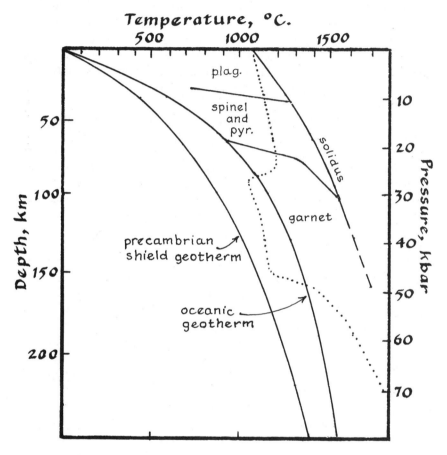

Fig. 1.2 Melting of mantle peridotite. The respective fields in which plagioclase, pyroxene, spinel, and garnet are stable are defined in relation to the geothermal gradients of the ocean, and Precambrian continental shields are shown for reference. The average temperature gradient for the earth as a whole would lie between these two curves. The temperatures at which mantle compositions begin to melt (called the *solidus*) are above the thermal gradients for dry conditions, but in the presence of small amounts of water *(dotted line)*, they intersect the oceanic geotherm at depths between 100 and 150 km, indicating that melting could occur there in the presence of small amounts of water. (Adapted from Green, D. H. 1970. *Trans Leicester Lit Phil Soc* 64:28–54; and Takahashi, E., and I. Kushiro. 1983. *Am Min* 68:859–79.)

rate of about 3 to 6° per kilometer, but if this gradient were to remain constant, almost all of the mantle would be far above its melting temperature, and as already noted, there is no evidence for large amounts of liquid at these depths. Various lines of geophysical evidence indicate that temperatures increase at a declining rate until the gradient becomes nearly constant at about a tenth of a degree per kilometer.

The only important part of the mantle where the geotherm is thought to approach temperatures at which melting begins is in a thin but extensive layer in the upper part of the mantle and at the base of the *lithosphere*. This zone lies at an average depth of 100 to 250 km under the continents but 70 km or so under the oceans and only a few tens of kilometers under oceanic ridges and other regions of intense volcanism. Seismic waves are slightly retarded in this *low-velocity zone*, indicating that temperatures at these depths are at least 1300°C. At low pressures, mantle rocks at this temperature would be largely molten, but because pressures raise their melting temperatures, the proportion of melt in the low-velocity zone is only about a percent or so. Thus, the low-velocity zone and the *asthenosphere* immediately below it seem to be favorable places for generating magmas, because it is in this region that the temperature gradient of the mantle is closest to the melting curve.

A number of observations support this interpretation. For example, seismic records at the Hawaiian Volcano Observatory indicate that the earthquakes that precede certain eruptions begin at a depth of about 60 km below Kilauea and rise steadily over months or years until lava finally appears at the surface. This observation was long interpreted as evidence that the magmas are produced at this or slightly deeper levels. Some magmas, however, must be formed at much greater depths for they bring up fragments, or "nodules," consisting of minerals that would be stable only at depths of 400 km or more, possibly as deep as the core–mantle boundary. On the other hand, other types of magma, particularly silica-rich varieties, may have much shallower origins, even in the crust at depths of only 20 to 30 km. Much depends on the composition of the source rocks and the process responsible for melting.

1.3 ▲ The Nature of the Earth's Mantle

The composition of the earth's interior can be deduced from a number of lines of evidence. The most useful of these is the information obtained from meteorites. These exotic rocks are fragments of other bodies in our solar system that broke up to form the asteroids and smaller fragments that occasionally intersect the earth's orbit. They can be divided into two broad types: iron meteorites, which are probably analogous to the earth's core, and stony meteorites, which consist mainly of silicate minerals similar to those of the earth's crust and upper mantle. Some of the most common types of stony meteorites, the *chondrites*, are thought to correspond to the original composition of the earth and other inner planets before formation of a core and crust. By a simple mass balance, one can arrive at the composition of the earth's mantle by subtracting from the average composition of these chondritic meteorites proportional amounts of the crust and core. The composition obtained in this way is close to that of a type of *peridotite* known as *lherzolite*. (See color illustration.) It is said to be "enriched" because in addition to the normal constituent minerals of lherzolites (olivine, clinopyroxene, and orthopyroxene), it contains small amounts of spinel or garnet and phlogopitic mica. Such rocks are found among the xenoliths, or "nodules," brought up in vents of deep-seated explosive eruptions, particularly diamond pipes. Moreover, rocks of this composition transmit seismic waves at velocities comparable to those measured through the mantle.

Large amounts of peridotite are exposed in the eroded cores of certain mountain ranges, such as the Alps and Sierra Nevada, where they have been brought to the surface by mountain-building forces that thrust part of the mantle over the crust. Unlike the enriched lherzolites, however, most of these are "depleted" peridotites (or harzburgites), consisting almost entirely of olivine and smaller amounts of pyroxene. Their average composition is chemically equivalent to the residue that would be left if a basaltic component were removed from an enriched peridotite. The proportions are approximately 1 part basalt and 3 parts depleted peridotite (Table 1.1). Thus, a basaltic magma could be derived from a spinel or garnet peridotite, leaving a residue consisting of only olivine and reduced amounts of pyroxene. Enriched and deplete mantle can be thought of as two end members between which there is a full range of intermediate compositions depending on the history of the rocks and the extent to which they have lost low-melting components.

Experimentalists have verified this hypothesis by determining the mineral assemblages that an enriched peridotite would have under mantle conditions and how these minerals would melt to form basaltic liquids. They find that with increasing pressure, the assemblage goes through the following transitions:

- Up to a pressure of about 10 kbars (roughly equivalent to a depth of 30 km), it consists of olivine, enstatite, clinopyroxene, and plagioclase.
- With a further increase of pressure to 20 to 30 kbars (a depth of 60 to 100 km), the composition is made up of olivine, aluminous

TABLE 1.1

Representative Chemical Compositions and CIPW Norms of an Average Chondrite, Mantle Peridotite, and "Primary" Basalts

	1	2	3	4	5	6	7
Weight Percent of Oxides							
SiO_2	48.0	45.16	44.59	48.28	45.70	42.00	50.53
TiO_2	0.13	0.71	0.06	0.22	0.05	4.23	1.56
Al_2O_3	3.0	3.54	2.98	4.91	1.60	12.50	15.27
Cr_2O_3	0.55	0.43	0.26	0.25	0.41	—	—
FeO*	13.0	8.45	8.34	9.95	5.90	12.90	10.46
MnO	0.4	0.14	0.17	0.14	0.09	0.19	—
MgO	31.0	37.47	41.10	32.53	42.80	9.59	7.47
CaO	2.3	3.08	2.22	2.99	0.70	11.13	11.49
Na_2O	1.1	0.57	0.22	0.66	0.09	2.47	2.62
K_2O	0.13	0.13	0.05	0.07	0.04	0.93	0.18
P_2O_5	0.34	0.06	0.01	—	0.01	0.66	0.13
Total	99.95	99.74	100.00	100.00	97.65	96.60	99.71
CIPW Normative Minerals							
Orthoclase	0.77	0.77	0.30	0.41	0.24	5.50	1.06
Albite	9.31	4.82	1.86	5.58	0.76	10.94	22.16
Anorthite	2.86	6.72	7.00	10.23	3.41	20.27	29.36
Nepheline	—	—	—	—	—	5.40	—
Clinopyroxene	5.21	6.47	3.11	3.64	—	25.13	22.01
Orthopyroxene	36.16	18.26	18.61	37.46	33.05	—	16.22
Olivine	41.17	58.89	66.85	39.76	57.92	16.98	4.50
Magnetite	3.14	2.04	2.02	2.41	1.43	3.12	2.12
Ilmenite	0.25	1.35	0.11	0.42	0.09	8.03	2.97
Apatite	0.74	0.13	0.02	—	0.02	1.44	0.30
Corundum	—	—	—	—	0.16	—	—

Note: Total iron oxides are expressed as FeO* (FeO + 0.9 Fe_2O_3). The normative minerals have been calculated assuming a ratio of $Fe^{3+}/Fe^{2+} + Fe^{3+}$ of 0.15. Note the abundance of normative olivine and pyroxene in rocks 1 through 5.

1. Composition of the earth's mantle calculated by subtracting from the composition of an average chondritic meteorite the composition of a core (rich in Fe, Ni, and FeS) having a mass equal to 32% of the earth.
2. Composition of mantle calculated as three-fourths peridotite and one-fourth basalt (Ringwood, 1966).
3. Peridotite inclusion in basalt, Itinome-gata (Kushiro and Kuno, 1963).
4. Spinel lherzolite inclusion in the tuffs of Salt Lake Crater, Hawaii (Kushiro, 1973).
5. Garnet lherzolite from the Wesselton diamond mine, South Africa (Mysen and Boettcher, 1975).
6. Alkali basalt with the characteristic composition of primary mantle-derived magma. Mururoa, French Polynesia (Bardintzeff, Demange, and Gachon, 1986).
7. Average composition of tholeiitic mid-ocean-ridge basalts (MORB).

pyroxenes, and spinel. In effect, plagioclase with its two end members, albite and anorthite, combines with olivine to form two of the components of pyroxene according to reactions such as the following:

$$NaAlSi_3O_8 \; + \; Mg_2SiO_4 \; \rightarrow \; NaAlSi_2O_6 \; + \; 2MgSiO_3 \qquad (1)$$
$$\text{albite} \qquad\quad \text{forsterite} \qquad \text{jadeite} \qquad\quad \text{enstatite}$$

$$CaAl_2Si_2O_8 \; + \; Mg_2SiO_4 \; \rightarrow \; CaAl_2SiO_6 \; + \; 2MgSiO_3 \qquad (2)$$
$$\text{anorthite} \qquad\quad \text{forsterite} \qquad \text{Tschermak's} \qquad \text{enstatite}$$
$$\text{molecule}$$

- At pressures greater than 20 to 30 kbars (depths of more than 100 km), the assemblage is that of a garnet lherzolite (olivine, enstatite, clinopyroxene, and pyrope garnet). The appearance of garnet is the result of chemical reactions, such as:

$$MgAl_2O_4 \; + \; 4 \, MgSiO_3 \; \rightarrow \; Mg_3Al_2Si_3O_{12} \; + \; Mg_2SiO_4 \qquad (3)$$
$$\text{spinel} \qquad\quad \text{enstatite} \qquad\qquad \text{pyrope} \qquad\qquad \text{forsterite}$$

$$m \, MgSiO_3 \cdot n \, MgAl_2SiO_6 \; \rightarrow \; Mg_3Al_{12}Si_3O_{12} \; +$$
$$\text{aluminous enstatite} \qquad\qquad\qquad \text{pyrope}$$

$$(m - 2) \, MgSiO_3 \cdot (n - 1) \, MgAl_2SiO_6 \qquad (4)$$
$$\text{aluminous enstatite}$$

Reaction (3) for the formation of pyrope garnet at the expense of spinel takes place between 21 kbars (at 1100°C) and 24 kbars (at 1300°C); reaction (4) occurs between 24 kbars (at 1300°C) and 31 kbars (at 1500°C).

These mineralogical relations hold only for volatile-free conditions, which are probably not entirely appropriate for the mantle. Small amounts of water, fluorine, and chlorine are present in hydrous minerals, such as amphibole or phlogopitic mica, and carbon dioxide can form carbonate minerals, such as dolomite. Thus, when experimentalists studied the same mineral assemblages in the presence of volatile components, they found that small amounts of other minerals are stable within certain ranges of pressure and temperature. Amphibole can be present up to pressures of about 30 kbars, beyond which it gives way to pyroxene, garnet, and water. Mica is stable throughout a wide range of pressures, provided that the rock contains appropriate amounts of water and potassium. Carbon dioxide combines with calcium and magnesium to form carbonates at pressures above about 20 to 25 kbars.

The lower mantle, at depths greater than 700 km, differs in that olivine is replaced by a mineral with the same chemical composition but with the more compact crystalline structure of perovskite. This gives the lower mantle a greater density that tends to prevent it from rising and mixing with the upper zone.

Although the composition of the mantle is much less complex than that of the crust, it would be an oversimplification to say that it is uniform. The convecting upper mantle is thought to have lost some of its lithophile and volatile elements that were extracted to produce the crust, atmosphere, and oceans. It overlies a deeper mantle that has not yet given up large amounts of these components. A seismic discontinuity marks the boundary between the two, and although heat is transferred from the lower to upper mantle, little exchange of chemical components takes place.

Several lines of evidence, such as varied mantle xenoliths and isotopic and trace-element differences of basalts, show that the sources of magma are far from homogeneous. Certain basalts, for example, seem to be derived from a mantle that was relatively enriched in aluminum, alkalies (K, Rb, and Cs), and the lighter rare-earth elements, whereas others come from sources that are relatively impoverished in these elements. It is thought that variations such as these reflect the different histories of certain parts of the earth's interior. If we could see the mantle, it might resemble a giant marble cake with one type of material forming swirls within another.

Some of these heterogeneities may be inherited from the earliest processes responsible for separation of the core, mantle, and crust, whereas others could be the result of continuing processes, such as the formation of oceanic lithosphere or subduction of crust into the mantle. Formation of some of these heterogeneities would require billions of years. At the same time, however, they would tend to be erased by mantle convection and, to a lesser degree, by diffusion that would take comparable times. Petrologists and geochemist have yet to reach a consensus as to what proportion of these variations are inherited from original inhomogeneities that still survive today and how much is the result of processes that have altered what was originally more homogeneous mantle. Most agree, however, that there is some kind of competition between the slow processes responsible for formation and destruction of these heterogeneities.

A complex model has been devised to explain the varied isotopic compositions of basalts (Fig. 1.3). It defines the following different end-member components of the mantle.

- A partly depleted mantle that is the source of mid-ocean-ridge basalts (MORB)

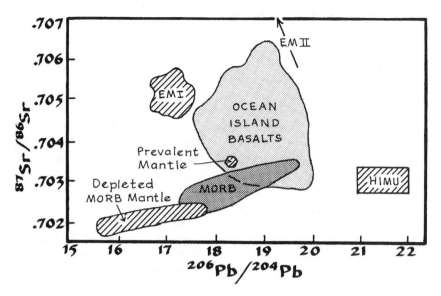

Fig. 1.3 Several mantle sources of basalt can be identified by their isotopic ratios of lead and strontium. Two of these (EMI and EMII) are more enriched in lithophile elements than in the "prevalent mantle"; the high-μ component (HIMU) is distinguished by an unusually large ratio of uranium to lead. Note that most ocean island basalts can be explained as combinations of different proportions of magmas from these three components. Mid-ocean-ridge basalts (MORB) are derived from mantle sources that have been depleted by one or more earlier melting events. (Adapted from Zindler, A., and S. Hart. 1986. *Ann Rev Earth Planet Sci* 14:493–571.)

- An enriched mantle (EMI) that may be a remnant of the early primitive mantle
- A second type of enriched mantle (EMII) that may be the result of subducted sediments
- A final type, HIMU, characterized by a high ratio (μ) of two isotopes, ^{238}U and ^{204}Pb, the origin of which is still not understood

Volcanic rocks with these features have been found only on a few islands, notably in Polynesia. It could be the result of fragments of subducted lithosphere from which certain mobile components have been lost.

The wide variety of basalts found in the oceans suggests that the mantle is made up of all four of these end members.

1.4 ▲ Mechanisms of Melting

Unlike simple substances such as ice, natural silicates do not melt at a fixed temperature but rather over a range between the *solidus*, where the first drop of liquid is formed and the *liquidus*, where the last crystal disappears. As natural silicates pass through this temperature interval, their physical properties, such as density and viscosity, change drastically. When the mantle begins to melt at its solidus, a thin film of liquid appears along grain boundaries and the partly melted rock still behaves like a solid. At least 2 or 3% melt must be present before it can be detected by seismic measurements.

If large amounts of magma are not normally present in the mantle, some special condition must be required to produce them. Melting cannot come about through a simple accumulation of heat produced by decay of radioactive elements; the rate at which heat is produced is much too slow to account for the rate at which magmas are produced. It could, however, result from one or more of several mechanisms that draw on the immense reservoir of heat stored in the earth's mantle (Fig. 1.4).

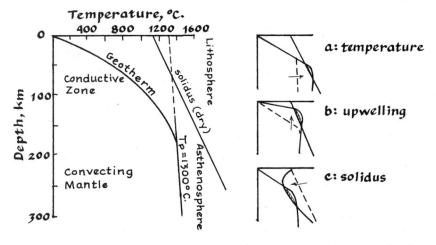

Fig. 1.4 The configuration shown on the left represents a possible relationship between a conductive thermal gradient in the lithosphere and an adiabatic gradient in a convecting asthenosphere having a potential temperature of 1300°C. These two gradients are separated by a thermal boundary layer near the boundary between the lithosphere and asthenosphere. The solidus temperature for the mantle is shown for dry conditions. The small diagrams on the right illustrate the effects of **(a)** raising temperature by an influx of heat, **(b)** lowering the pressure by upwelling of the asthenosphere, and **(c)** lowering the solidus curve by introducing another component, such as volatiles.

For volatile-free compositions, the curve for the solidus temperature of the mantle (the temperature at which peridotites begin to melt) increases at a nearly linear rate of about 3°C per kilometer from about 1200°C at the surface to 1400 to 1500°C in the low-velocity zone where it approaches or intersects the geotherm. At these depths, melting could be caused by one or more of three basic mechanisms illustrated in Fig. 1.4: (a) an influx of heat and increase of temperature at constant pressure and composition, (b) a decrease of pressure at constant heat content and composition, or (c) a change of composition at constant heat content and composition. We can consider each of these independently.

(a) If an influx of heat from some deeper source enters a horizon where the mantle is already close to the solidus, temperatures could rise until melting begins. Alternatively, if rocks are depressed to deeper levels of higher temperature, they will gain heat from their surroundings. In either case, temperatures could rise until they reach the solidus, where melting begins. Thereafter, most of the added heat would be absorbed in melting, and temperatures would rise much more slowly with each joule of additional heat producing about 0.25% melt. A mechanism of this kind is probably responsible for generation of siliceous magmas, such as the granitic plutons of mountain ranges like the Alps. Collision of continental plates thickens the crust, and as the roots of mountains are depressed to deeper, hotter levels, their temperatures rise until melting begins.

(b) If, instead of an influx of heat, the rocks are raised to a level of lower pressure, possibly as a result of convective upwelling of the mantle, rocks that were already close to their solidus temperature would now be in a realm where they would be partly liquid. Even though no energy is added, crystals would begin to melt by drawing on heat stored in the rocks. Raising a mantle rock 1 km lowers its solidus temperature about 3°C. At the same time, the adiabatic expansion from relief of pressure lowers the temperature of peridotite about 0.3° per kilometer. If the specific heat of the rocks is about 1.3 joules $g^{-1} deg^{-1}$, the amount of heat available for melting would therefore be $(3 - 0.3) \times 1.3 = 3.5$ joules per gram, or enough to melt nearly 1% of a rock with a heat of fusion of 400 joules per gram. A rising mantle with an average temperature of 1340°C would begin to melt when it reaches a depth of 50 km and would be 25% liquid if it reached the surface. The large amounts of magma produced at hotspots and oceanic spreading ridge could be generated easily by this type of "decompression melting."

(c) A third melting mechanism requires neither an influx of heat nor a change of pressure but rather an addition of some component, such as H_2O, that lowers the melting temperature. This fluxing action might result from an introduction of volatiles, either from a source at greater depths or from breakdown of a volatile-bearing mineral in the rock itself. Although the combined contents of H_2O and CO_2 in basalts are probably less than 1 weight percent, even these small amounts could have a large effect on melting. Melting temperatures of basalt are lowered as much as 100°C by only 0.1% water; other volatile components, as well as alkalies, have comparable effects.

The magmas erupted from volcanoes above subduction zones are thought to be produced in this way. As a slab of oceanic lithosphere descends into the mantle, the crustal rocks are dehydrated, and their volatile components rise into the overlying mantle rocks, lower their melting temperature, and induce melting.

1.5 ▲ Ascent of Magmas from Their Source

The manner in which magmas rise and eventually reach the surface can only be inferred from a few tenuous lines of indirect evidence. Although geologists may differ in their interpretations of the exact mechanisms, all agree that the primary factors are the physical properties of the magmas, chiefly density, viscosity, and heat content.

The densities of primary basaltic magmas are about 2.8 to 2.9, substantially less than that of the mantle from which they are derived (about 3.3). They may rise, either by dispersed porous flow like water in an aquifer or as a distinct bulbous mass moving through the plastic mantle, much in the way diapirs of salt domes rise through weak sediments. Because the mantle minerals that contribute the most to basaltic liquids are the densest components of the original peridotite, the refractory minerals that remain after partial melting are probably lighter than the original mantle. This is especially true when melting takes place at depths where the source rocks contain garnet. Thus, both the liquid and refractory residue are buoyant with respect to the original fertile mantle and may rise together as a unit. As they do so, the melt will continue to re-equilibrate until the liquid and crystalline fractions finally separate at some shallower depth. During this early stage, volatiles probably remain in solution and have little effect on the rate of ascent until the magma reaches shallow depths.

At higher levels where the rocks of the lithosphere are more rigid and fracture under stress, magma can no longer rise by plastic flow. In tensional regimes, such as oceanic spreading axes, dilational stresses may open fractures, allowing magmas to continue their ascent. Elsewhere, pockets of magma may migrate upward by a process known as "stoping," in which pieces of the roof break away and sink or become assimilated. In either case, the ability of magma to rise through the lithosphere is severely restricted, especially in the crust, where the density and temperature of the rocks may be much less than that of the magma.

Melts at their source in the mantle are at temperatures that would be well above their liquidus as low pressure, and yet magma rarely arrives at the surface in a superheated condition, that is, at temperatures above their liquidus. As magma rises, it cools as a result of two effects. Just as air cools as it rises and expands, magma cools *adiabatically* as it rises to levels of lower pressure; that is, its temperature declines even if no heat is lost. The temperature, T (in degrees Kelvin), of the magma varies with pressure, P, according to the following relation:

$$\frac{dT}{dP} = \frac{T\alpha}{41.84\ \rho\ C_P}$$

where α is the coefficient of thermal expansion (about 3×10^{-5} deg^{-1}), ρ is density (about 2.8 g cm^{-3}), and C_P is heat capacity (about 0.3 cal g^{-1} deg^{-1}). The factor 41.84 converts heat to its mechanical equivalent. Thus, for mantle temperatures and densities, the effect of adiabatic cooling would be as follows:

$$\frac{(1200 + 273) \times 3 \times 10^{-5}}{41.84 \times 2.8 \times 0.3} = 0.001° \text{ bar}^{-1}$$

or between 0.2 and 0.3° per kilometer of rise. The slope of the melting temperature is about 3° per kilometer, so adiabatic cooling has a relatively minor effect. More important is the loss of heat to the cooler rocks through which the magma rises. If the temperature of the melt exceeds that of its surroundings, minerals of the wall rocks may begin to melt and the large amounts of heat absorbed in this way quickly bring temperatures down to the melting curve. In the crust, this effect may be so great that the magma will begin to solidify and be too viscous to continue its ascent.

A further limitation is the density contrast between magma and overlying rocks. If the crustal rocks are less dense than the magma, the gravitational driving force is diminished and the height to which magma can rise is severely limited. This is not to say that magma cannot rise through lighter rocks; as long as the total pressure on the source exceeds the hydrostatic pressure of the column of magma, the magma can continue to rise. Nevertheless, the magma begins to lose its buoyancy when it passes from the dense mantle into lighter rocks of the crust, and when this happens, it may be injected into fractures in the crust where it begins to cool and crystallize. If the influx continues, each addition of new magma supplies new heat and retards solidification, so magma may accumulate in a large, persistent reservoir.

Shallow reservoirs of this kind are thought to underlie large volcanoes that remain active for many thousands of years. The volumes of some of these "magma chambers" are probably tens or even hundreds of cubic kilometers.

When magma resides in such a chamber for centuries, it can undergo important compositional changes before finally resuming its course toward the surface. For example, as crystallization advances, the density of the remaining liquid may decline and volatile components may approach saturation in the reduced volume of liquid. As a result, the magma may become buoyant and renew its ascent.

In certain types of large-scale magmatism in which channels provide easy access to the surface, it is possible that magma may rise directly from the mantle without undergoing major compositional changes. For example, volcanoes such as the Icelandic volcano Krafla that commonly erupt large amounts of lava of very uniform composition are thought to be fed by magma that has scarcely paused in its rise directly from its mantle source. This is exceptional, however; most lavas have compositions unlike those of primary melts of the mantle.

1.6 ▲ Principal Types of Magmas

Most magmas have compositions that are distinctive of the tectonic setting in which they are found. Andesites, for example, are the characteristic rock of subduction-related volcanoes. One would be very suspicious if such a rock were reported from an oceanic island, such as Hawaii. Similarly, granitic plutons and large rhyolitic ignimbrites are confined exclusively to continental settings and have never been found in the oceans.

Because magmas have chemical compositions markedly different from that of any reasonable mantle rock, they cannot be products of the total melting of their source rock; they must be the result of different degrees of partial melting. At the very earliest stages of melting, certain volatile and easily fusible components, such as the alkalies, contribute disproportionately to the liquid. Elements, such as Zr, Hf, U, and Ba, are said to be *incompatible* because they are almost totally excluded from most common minerals and enter readily into the very first products of melting. With more advanced melting, the concentrations of these elements are diluted as larger amounts of the more refractory components contribute to an increasing amount of liquid.

The maximum amount of melting indicated by modern volcanic rocks is seldom more than 20 to 30%. Certain ancient lavas, such as the very magnesium-rich *komatiites*, are thought to have been products of at least 40% melting, but magmas of this kind were common only during Precambrian time when the earth's thermal gradient was much steeper and tectonic processes more active. They are very rare in modern volcanoes.

Melting of the crust, or *anatexis*, produces magmas rich in silica, alumina, and alkalies, especially potassium. It is thought to have been responsible for the silica-rich volcanic and plutonic magmas that could not be derived in such volumes from a source of peridotitic composition. Melting of continental crust is confined to two types of conditions. First, when dense mafic magmas reach the base of lighter continental crust, they tend to collect there and raise the temperature of the overlying rocks until they begin to melt. The voluminous rhyolitic magmas erupted from Yellowstone are thought to originate in this way (see Chapters 8 and 10), and the sillimanite-bearing ignimbrites of Peru and Chile have been attributed to melting under similar conditions in the roots of the Andes. Second, in regions of plate collision, depression of the continental crust to levels of elevated temperature may cause melting. The Tertiary cordierite dacites and tourmaline rhyolites of Tuscany are products of the collision of the Adriatic and Eurasian plates.

A melt that comes directly from its source without undergoing any subsequent change of composition, either by crystallization or contamination, is referred to as a *primary magma*. As a rule, few magmas reach the surface in their original, unaltered state. With small amounts of crystallization, they are quickly depleted of certain elements, such as Mg, Ni, and Cr, that enter early crystallizing minerals. At the same time, they may be contaminated with crustal material picked up as they rise toward the surface.

Even when these effects are discounted, primary magmas have subtle but important differences from one tectonic setting to another. The variations reflect the nature of the source rocks and conditions, such as temperature, pressure, and the concentrations of volatiles that prevail where melting takes place. Experimental studies have shown how these factors can account for the main types of primary magmas that characterize the principal tectonic settings in which volcanism occurs.

1.7 ▲ Magma Series

Although names have been coined for a vast number of igneous rocks, the great majority fall into a small number of major genetic types. These groupings were initially set up by petrologists who noted compositional features shared by the suites of rocks of a particular locality. Even though the lavas of a certain volcanic center may not be uniform, they form gradational series that share certain mineralogical or geochemical characteristics. For our present purposes, we shall concentrate only on the most common types and ignore the countless subdivisions set up for more specialized purposes.

With the exception of rare carbonatites (see Chapter 5), all magmas are essentially molten silicates with lesser amounts of Al, Fe, Mn, Mg, Ti, Ca, Na, K, P, and a host of trace elements (Table 1.1). They can be classified as either *subalkaline* or *alkaline,* depending on their proportions of silica relative to other components. Alkaline rocks, as their name implies, contain more alkalies as well as magnesia and iron but relatively less silica. They can crystallize a feldspathoid, such as nepheline ($NaAlSi_2O_6$), in place of plagioclase ($NaAlSi_3O_8$) and olivine ($MgSiO_4$) in place of hypersthene ($Mg_2Si_2O_6$). These minerals need not be physically present in the actual or *modal* assemblage, especially if the rock cooled quickly and contains appreciable amounts of glass. Ideally, however, they would be among the theoretical or *normative* minerals making up an equilibrium assemblage.

The compositional differences between alkaline and subalkaline basalts may seem trivial, but these small differences are magnified manyfold as the compositions of derivative magmas evolve. The end product of subalkaline differentiation is a quartz-rich rock, such as rhyolite or granite; that of alkaline magmas is phonolite or syenite with compositions much richer in alkalies and poorer in silica (Fig. 1.5a). As in all natural classification systems, some individual rocks or series of rocks straddle the boundaries between subdivisions. Some magmas that are only mildly deficient in silica differentiate toward end products, such as trachytes, that are intermediate between rhyolites and phonolites in the sense that they may contain no feldspathoid but they also lack appreciable amounts of quartz.

Alkaline and subalkaline magmas are found in almost all geological settings but tend to be characteristic of certain tectonic regimes. The great volcanoes of Hawaii and Reunion, for example, erupt alkaline basalts in their earliest stages but soon change to subalkaline during most of their periods of active growth; in their waning stages, these volcanoes produce small amounts of increasingly alkaline magmas. This change tells us that the conditions governing the generation of magmas in the mantle are changing with time and enables us to interpret processes deep in the earth beneath the volcanoes.

Each of these two broad groups has major subdivisions. *Tholeiites* are a variety of subalkaline rocks characterized by strong enrichment of iron in the early and middle stages of differentiation, but in later stages, they become much richer in silica and may evolve to rhyolite or granite. This is in contrast to *calc-alkaline* series that tend to have a more-or-less linear variation between their most primitive and their

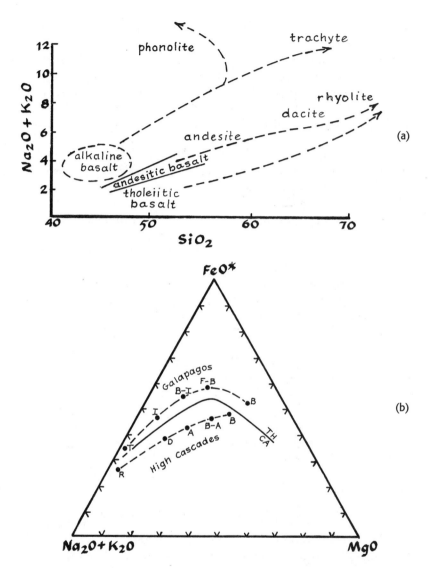

Fig. 1.5 (a) Alkali–silica diagram showing division between tholeiitic, calc-alkaline, and alkaline series. **(b)** The tholeiitic (TH) and calc-alkaline (CA) magmatic series can easily be distinguished by the differing proportions of iron in their intermediate members. The solid line shown on this diagram for iron, alkalies, and magnesia is an empirical division between the two series. Iron is taken as FeO * (FeO + 0.9 Fe$_2$O$_3$). The Galapagos lavas are an excellent example of a differentiated tholeiitic series. Starting with basalt (B), they evolve through ferrobasalt (F-B), basic icelandite (B-I), and icelandite (I) to trachyte (T). The rocks of the High Cascades illustrate the calc-alkaline trend from basalt (B) through basaltic andesite (B-A), andesite (A) and dacite (D) to rhyolite (R).

most silica-rich end members (Fig. 1.5b). Although the tholeiitic and
calc-alkaline series both begin with basalt and end with rhyolite, their
intermediate compositions are very different. The names of their
intermediate members reflect their characteristic occurrences: calc-
alkaline *andesites* are the dominant lavas of the great volcanoes of the
Andes and other parts of the Circum-Pacific region; tholeiitic *ice-
landites* are found chiefly in the large volcanoes of oceanic islands like
Iceland.

The other major group, the alkaline rocks, has a bewildering variety
of types, each distinguished by distinctive mineralogical or chemical
features. The dominant alkali may be either sodium or potassium, and
the degree of silica deficiency may be slight or extreme. Potassic varieties are confined almost exclusively to the continents and are particularly characteristic of volcanoes behind the major calc-alkaline
chains of continental margins and island arcs. Sodic varieties are
found in the volcanoes of both continents and oceans. In some regions, such as the great Rift of East Africa, highly sodic lavas are
erupted from some volcanoes, while highly potassic lavas are produced from a nearby neighbor.

1.8 ▲ Characteristics of Volcanism in Different Settings

Each of these main types of magma tends to occur in a distinct tectonic setting reflecting the manner in which they are generated and
rise through the lithosphere. These associations are found throughout
the world and appear to have prevailed through most of geological
time. Thus, we can take the following as general rules:

- At oceanic spreading axes, the characteristic lavas are weakly
  differentiated tholeiites. Alkaline rocks are rare at oceanic ridges
  but are common in continental rift zones.
- In subduction-related volcanic belts, calc-alkaline (andesitic)
  lavas are by far the dominant species, especially along continental margins. Tholeiitic basalts are found in many island arcs, but
  they evolve toward andesites very similar to those of normal
  calc-alkaline series.
- In zones of continental collision, volcanism is rare and widely
  scattered. The rocks tend to be potassium-rich calc-alkaline
  types, but alkaline and transitional rocks are almost as
  common.
- In intraplate regions of the oceans, alkaline rocks are very common but they are often preceded by tholeiitic series. Alkaline

rocks are also common in the interior regions of continents, but so too are tholeiites, especially where magma rises through great thicknesses of granitic crust.

- Large-scale melting of the continental crust *(anatexis)* in zones of subduction or intraplate settings produces voluminous siliceous magmas of calc-alkaline character.

Each of these settings has a distinctive mechanism of magma generation. At oceanic spreading axes, the low-velocity zone rises to shallow depths where the pressure under ridges is relatively low, probably less than 20 kbars. The dominant mechanism of magma generation is probably decompression melting. Experimental studies indicate that MORB of tholeiitic composition are produced by about 20 to 30% melting of a mantle that is relatively poor in incompatible elements.

The calc-alkaline basalts and andesites of subduction regimes are produced in a complex structural environment in which thermal and chemical conditions are drastically altered by the subducted oceanic crust. Fluids released from the descending slab must play an important role by inducing melting of the overlying mantle wedge.

Most basalts of intraplate regions of either continental or oceanic character are generated at depths and pressures (>20 kbars) where the thermal gradients are abnormally steep. The degree of melting at these hotspots varies widely. The first few percent of melting yields alkaline basalts, but when melting reaches about 20 to 30%, the magmas become tholeiitic. Although various types of magma are found on oceanic islands, most can be classified as either an alkaline variety known as *ocean island basalt (OIB)* or as *ocean island tholeiites (OIT)*. The differences probably result from the nature of their source rocks.

Thus, the character of any given magma depends on tectonic and thermodynamic conditions and on the degree of melting of a heterogeneous mantle source. The basalts of oceanic islands could come from ancient fragments of subducted oceanic lithosphere stored in the lower part of the upper mantle or possibly as deep as the core–mantle boundary. After residing at these depths for millions or even billions of years, this material may rise in the form of diapirs and be partially melted at shallower depths of the mantle. Another hypothesis holds that these fragments remain scattered at different depths in the upper mantle before rising and melting. Whatever the exact mechanism may be, the heterogeneities inherited from the ancient subducted lithosphere seem to be closely associated with hotspots.

In Chapter 2 we shall see how these magmas evolve by shallow differentiation to produce a diverse variety of lavas and plutonic rocks.

Suggested Reading

Green, D. H., and A. E. Ringwood. 1967. The genesis of basaltic magmas. *Contrib Mineral Petrol* 15:103–90.
Although somewhat outdated, this is an important contribution to our understanding of the origins of magmas.

McBirney, A. R. 1993. *Igneous petrology,* 2d ed. Boston: Jones & Bartlett.
A comprehensive textbook on the nature, occurrence, and origin of igneous rocks.

Yoder, H. S., Jr. 1976. *Generation of basaltic magma.* Washington, DC: National Academy of Sciences, 265 p.
An advanced treatment of the origins of basaltic magmas with emphasis on experimental studies.

Zindler, A., and S. R. Hart. 1986. Chemical geodynamics. *Ann Rev Earth Planet Sci* 14:493–571.
An advanced treatment of the isotopic systems of the mantle and their use in determining the origins of basaltic magmas.

Storage and Differentiation of Magmas

▼

2.1 ▲ Subvolcanic Reservoirs

Few magmas erupt at the surface without first collecting in a shallow crustal reservoir. Although the evidence for these storage chambers under active volcanoes is largely indirect, certain of their features can be deduced from geophysical measurements and studies of deeply eroded volcanoes.

Because the velocities of seismic waves passing through partly molten rocks are measurably slower than those passing through normal crust, it is possible to detect the presence of bodies of magma by measuring the retardation of waves following different paths through the crust beneath a volcano (Fig. 2.1a). This technique, known as *seismic tomography,* has been used to define the size and location of magmatic reservoirs under exceptionally large volcanic centers, such as Yellowstone and Long Valley, California, but it has had limited success in detecting magma under smaller volcanoes. Only bodies with dimensions substantially greater that the wave length of tectonic earthquakes (about 1 km or more) can be detected in this way.

Another seismic technique (Fig. 2.1b) has been used to surmount this limitation. Because the shear strength of rocks is very sensitive to temperature, small stresses in hot rocks are relieved by gradual deformation rather than by sudden brittle failure. For this reason, there can be a deficiency of small earthquakes in parts of the crust with abnormal temperatures, even where only small amounts of magma are present. With an array of sensitive seismometers, a record of earthquakes can be compiled to show the distribution of shocks of a given

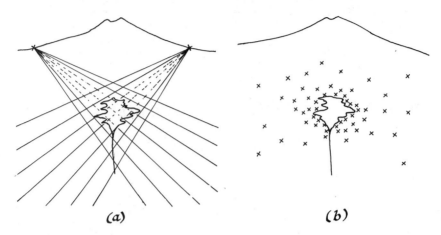

Fig. 2.1 Magma chambers can be detected by the retardation of seismic velocities of waves following different paths from distant sources **(a)** or by the distribution of small earthquakes **(b).** The first of these methods has been used successfully to detect a large reservoir of magma under the Yellowstone caldera; the second has enabled volcanologists to outline a body of magma under Mt. St. Helens.

magnitude. For example, a relative deficiency of earthquakes with a magnitude less than 1.0 has been used successfully to define the depth and size of small bodies of magma beneath composite volcanoes, such as Mt. St. Helens.

Elevated temperatures also affect the electrical conductivity of rocks. Conductivity increases rapidly at temperatures approaching the melting range. By inducing a strong current between two electrodes and measuring the resistivity of the crust, one can detect thermal anomalies to depths of several kilometers. This technique has been used successfully to explore geothermal resources. A similar technique uses the effects of magnetic "storms" to detect anomalous zones within the crust. These magnetotelluric techniques have revealed bodies of magma beneath some of the volcanoes of the Cascade Range.

Shallow intrusions of magma can be detected from the swelling that they produce at the surface. The best example is probably the large volcano Kilauea on the island of Hawaii where the entire summit region is observed to rise when new magma inflates the edifice and then to subside again after a discharge of lava, usually on the flanks. Using the geometrical form of the swelling, one can estimate the size and depth of the intrusion.

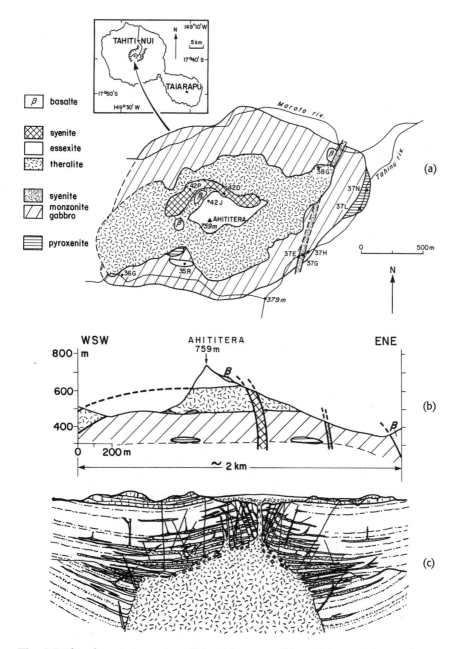

Fig. 2.2 The plutonic intrusion of the Tahiti-nui caldera. **(a)** Simplified geologic map. The inset above the map **(b)** is a schematic section showing a possible interpretation. The horizontal and vertical scales are the same as that of the map in **(a)**. More typical of the magma bodies under large continental volcanoes and calderas is the Tatoosh pluton under Mt. Rainier **(c)**. (**a** and **b**, From Bardintzeff, J. M., et al. 1988. *J Volc Geoth Res* 35:31–53, **c,** From Fiske, R. S., C. A. Hopson, and A. C. Waters. 1963. *U S Geol Surv Prof Paper* 444, p. 93.)

One would think that deep erosion would reveal the magma chambers that fed ancient volcanoes, but evidence of this kind is difficult to interpret because erosion to such levels normally removes the entire overlying volcano and makes it difficult to relate the intrusive rocks to the structure of a volcano that has long since disappeared. It is particularly difficult to determine whether these intrusions fed surface eruptions. For example, moderately large, coarse-grained intrusions can be seen in the cores of some of the Tertiary volcanoes of the Cascade Range, but it is seldom possible to link them directly to surface eruptions. They may merely have intruded the base of the volcano after most eruptive activity had come to an end.

Subvolcanic plutons differ widely in size and shape according to the structural setting into which they were intruded. Some are broad, flat lenses (Fig. 2.2a and b); others are steep-sided stocks that have stoped their way upward through the crust (Fig. 2.2c). Their shapes and sizes depend chiefly on the state of stress and physical properties of the crustal rocks they invade. In extensional settings where the least principal stress is horizontal, magma can force aside its walls and form dikes. If the least stress is vertical, the magma lifts its roof and intrudes as a sill. Where the horizontal and vertical stresses are about equal, it may rise as a pipe-like body or along ring dikes surrounding a central subsiding block (Fig. 2.3).

The magma reservoirs under large active volcanoes, such as Kilauea, seem to be swarms of dikes and sills maintained by periodic injections of magma coming from the mantle. When eruptions drain part of the volume, the loss is balanced by an influx of new "primitive" magma. Under such conditions, the composition and physical properties of the magma can remain more or less constant through long periods of activity. As the influx of new magma declines, however, eruptions are separated by longer intervals of repose during which the magma has time to crystallize and evolve compositionally. The eruptive products then acquire a wider range of chemical and physical properties.

Few magmas discharged after long periods of repose retain primitive mantle compositions. Quite often, the first magma to appear in an eruption is more evolved and gas-rich, whereas the last to emerge tends to be more basic and crystal-rich. Variations of this kind result from slow processes that accompany the prolonged cooling of large intrusions. As magma cools, successive minerals begin to crystallize, and this in turn affects the composition and physical properties of the remaining liquid. Because the behavior of volcanoes is so closely linked to these compositional differences, it is important to understand the processes by which magmas evolve.

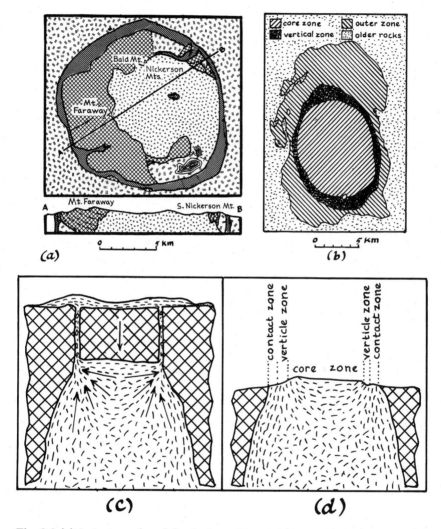

Fig. 2.3 (a) A ring complex of the Ossippee Mountains, New Hampshire, was emplaced during the Devonian period. **(b)** The El Pinal tonalite of Baja California was a similar body underlying a caldera but has been eroded to a greater depth. The flow texture preserved in the coarse-grained rocks shows where the magma fed a ring fracture surrounding a subsiding central block (**c** and **d**). (**a**, After Kingsley. 1931. *Am J Sci* 22:139–68.).

2.2 ▲ Some Basic Principles of Magmatic Differentiation

Most of the compositional diversity of igneous rocks can be explained in terms of one or more of four basic processes: *partial melting, assimilation, mixing,* and *crystal fractionation.* As explained in Chapter 1, the first of these operates where magma is generated in the mantle. The second takes place during its ascent through the mantle and crust, and the last two prevail while the magma resides in a subvolcanic reservoir. As we saw in the opening chapter, some of the compositional variations of magmas reflect the conditions under which they are generated, particularly the degree and depths of melting in the mantle. The resulting variations consist mainly of small differences in the concentrations of alkalies, magnesium, and certain trace elements.

2.3 ▲ Assimilation

Compositions can also be altered by assimilation of the rocks of the mantle and crust through which magmas rise. The most common sign of contamination is exotic crystals, known as *xenocrysts,* that are not in equilibrium with each other or with the liquid in which they reside. Incompatible assemblages, such as quartz and forsteritic olivine, may be found together in basalts that have been contaminated by granites, sandstones, or other types of quartz-bearing crustal rocks. As this foreign material is digested, it contributes distinctive components to the magma. Certain *lithophile elements,* such as potassium and uranium, or radiogenic isotopes of strontium or lead, which have greater concentrations in the continental crust than in mantle-derived melts, provide a way of detecting even small amounts of contamination with crustal material.

2.4 ▲ Mixing and Mingling of Magmas

Another form of assimilation results when two magmas of differing origins combine to produce a hybrid composition. A distinction is often made between "mixing" and "mingling" of magmas, with the former term being applied to a process in which the end product is homogenized and the latter to one in which the two components are still distinguishable. In the case of mingling, the identities of the components are easily recognized, but mixing produces a product that is more difficult to interpret.

The mixing of two primary magmas was first proposed as an explanation for the compositional variations of igneous rocks by R. W. von

Bunsen, who, in 1851, noted the bimodality of the lavas of Iceland. Basalts and rhyolites make up 85 and 12%, respectively, of the total volume, and intermediate rocks only about 3%. Von Bunsen proposed that the latter were the result of mixing two primary magmas, one basaltic and the other rhyolitic. Similar relations have since been found in many parts of the world. One of the most extreme examples is the volcanic complex of Medicine Lake, California, where a primary basaltic magma with 48.8% SiO_2 has mixed with a rhyolite with 73.2% SiO_2 to produce a wide range of intermediate products.

The two components of mixtures may not necessarily be independently derived primary magmas; in many, if not most, cases, they are genetically related, either through differentiation or assimilation. A fresh, primitive magma may enter a magma reservoir and mix with an older magma that has evolved from a similar parent during a long period of repose. The result is a magma of intermediate composition.

2.5 ▲ Crystal Fractionation

By far, the most important process affecting the compositions of magmas is crystal fractionation. Although it may involve a variety of physical mechanisms, most of the compositional differences observed in the eruptive products of volcanoes can be attributed to the changes that result as magmas slowly crystallize en route to the surface or in a subvolcanic reservoir.

Two eruptions of the Hawaian volcano Kilauea, one in 1959 and the other in 1965, provided an opportunity to observe the compositional evolution of magma directly. In both instances, basaltic lava flowed into a pit–crater where it ponded to depths as great as 100 m and slowly cooled and crystallized. The rates of solidification, although rapid compared with those of plutonic rocks, were slow enough to permit geologists to drill through the crust and follow solidification of the lava as its temperature slowly declined.

The crystallization of these basalts illustrates some of the properties of natural magmas and how they differ from most other liquids (Fig. 2.4). Crystallization did not occur at a single temperature, as ice does when it freezes at 0°C. Instead, it extended through at least 200°C of cooling as a succession of minerals nucleated and grew with falling temperatures. Augite began to crystallize between 1180 and 1160°C, then plagioclase appeared at around 1160°C, and iron oxide at 1065°C. As these minerals crystallized, the liquid changed composition—first becoming more iron-rich until iron-oxide minerals appeared, then becoming poorer in iron but richer in silica, alkalies, and other elements excluded from the crystallizing minerals.

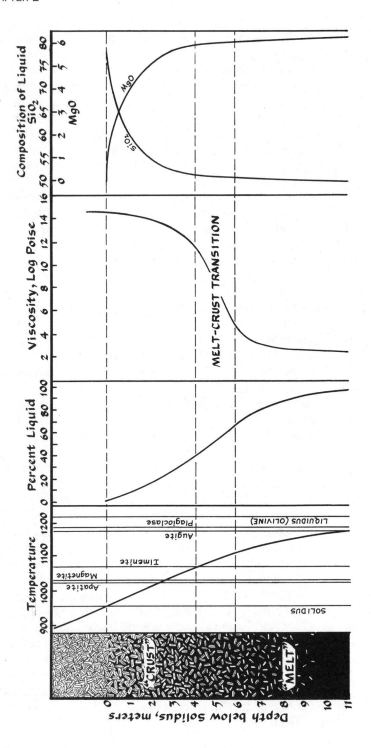

Fig. 2.4 As the basaltic magma in the lava lake of Kilauea Iki, Hawaii, solidified, the concentrations of the various components changed in response to progressive crystallization. The plot above shows the variations of several of these components in liquids (quenched to glass) in samples recovered by drilling into the crystallizing basalt. Note that some components are depleted from the liquid, while others are enriched. The rates of enrichment or depletion depend on the compositions and proportions of crystallizing minerals. If the differentiated liquids were removed and allowed to crystallize elsewhere, they would form a series of rocks of different compositions. Without this second step of segregation, the process of differentiation is incomplete. (Data from W. C. Luth, Sandia Corporation, personal communication.)

The final product of this process differed markedly in composition from the initial lava. In contrast to the original basaltic composition with about 50% SiO_2, the last liquid to crystallize contained more than 75% SiO_2 and had a composition close to that of rhyolite. Had the liquid fraction been separated at any intermediate stage, it could have formed a rock with a composition very different from that of the original lava.

Magmas crystallize over wide ranges of temperature because the composition of the liquid is constantly changing as crystallization advances. The change comes about because the compositions of the growing minerals differ from that of the liquid from which they form. Note in Fig. 2.5 that crystallization begins at a temperature that is a function of composition (as well as pressure and other thermodynamic conditions) and that the composition of the solid precipitated at any temperature differs from that of the liquid. The domain between the beginning of crystallization (or *liquidus*) and complete solidification at the *solidus* is one in which the magma consists of both liquid and crystals.

Nearly all magmas erupt at temperatures at or slightly below their liquidus. Most contain visible, well-formed crystals, known as *phenocrysts,* that nucleated and grew slowly at moderate depths, probably while the magma was rising through cooler rocks or residing at an intermediate level a few kilometers below the surface. These phenocrysts, some of which reach 2 or 3 cm, are larger than the tiny *microlites* of the matrix that crystallize during a later stage of more rapid cooling and crystallization after the magma has reached the surface. Some of the remaining liquid is quenched to a glass that, together with the microlites, forms the *groundmass* of the rock.

As crystals grow and extract their essential elements from the liquid, these components are increasingly impoverished in the remaining magma. As a result, the composition diverges more and more from that of the original magma. While *included elements* that enter in large amounts into the crystallizing minerals are depleted, the *excluded elements* that are rejected by growing crystals are residually enriched as they are concentrated in a diminishing volume of liquid. This basic principle governs most of the compositional evolution or differentiation of magmas.

Thus, two factors govern the compositions of differentiated magmas. The order of appearance and compositions of the crystallizing minerals determine the *trend* of compositions, while the amount of crystal fractionation determines the *degree* of differentiation. The initial composition of the primary basalt and the conditions under which it crystallizes define the chemical trend, but the efficiency of the physical processes of differentiation determines how far the composition will evolve down its liquid line of descent. Depending on the initial composition of the primary magma and the nature and proportions of minerals it crystallizes, the course of changing compositions differs from one type of magma to

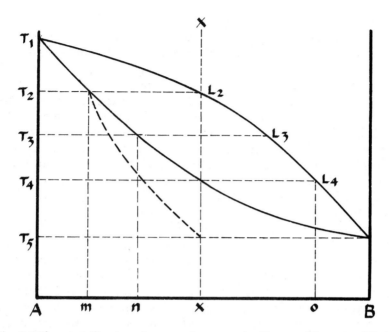

Fig. 2.5 The crystallization of most magmas can be illustrated by a simplified diagram for two components forming a continuous series over a range of temperatures. The upper curve (the *liquidus*) defines the temperature of crystallization as a function of composition defined by two end members, A and B; the lower curve (the *solidus*) defines the composition of crystals precipitated at that same temperature. Because the crystals are richer in the high-temperature end-member A than the liquid at the same temperature, the remaining liquid becomes residually enriched in B as temperature falls and more crystals are removed. A liquid of composition X, initially at temperature T_1, would begin to form crystals of composition m on reaching temperature T_2. As temperature declines, the compositions of both liquid and solid evolve along the liquidus and solidus. If the first formed crystals re-equilibrate perfectly during this cooling, the last drop of liquid with composition L_4 disappears at temperature T_4. If they are separated from the liquid and retain their original compositions, the bulk composition of the crystals follows a path such as that shown by the dashed curve and the liquid evolves to the end point, B, at T_5.

another. In this way, the various types of magma series mentioned in Chapter 1 evolve along distinct lines of evolution.

2.6 ▲ Physical Properties

An important consequence of magmatic differentiation is its effect on physical properties, such as density and viscosity. The density of a magma (Figs. 2.6 and 2.7) may either decrease or increase with differentiation, depending on the compositions of the liquid and crystallizing minerals (see Fig. 2.6). Most extreme differentiates are less dense than the primitive magmas from which they have evolved. Rhyolitic melts, for example, have a density of only about 2.4 g/cm³, whereas the corresponding value for basalts is at least 2.7 g/cm³. This difference is the result of removal of heavy components, such as Fe and Ti, in olivine, pyroxene, and oxide minerals. Dissolved volatile components, which become increasingly concentrated in the diminishing volume of remaining liquid, may reach the point of saturation and begin to form bubbles. When this happens, the magma expands and its density declines abruptly. In extreme cases, such as rhyolitic pumice, it may even become lighter than water.

Depending on their composition, temperature, and crystal content, magmas can have an enormous range of viscosities (Fig. 2.7a). In general, viscosity increases with declining temperature and increasing silica content, but it is reduced by other components, such as alkalies and dissolved volatiles, particularly water, chlorine, and fluorine. Thus, rhyolitic melts are normally more viscous than basalts, mainly because they are richer in silica and their temperatures are much lower; but if they contain large amounts of dissolved volatiles and are relatively rich in alkalies, they can be almost as fluid as basalts.

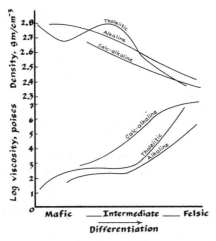

Fig. 2.6 Variations of density and viscosity of the three main types of magma series as they evolve toward more felsic compositions. The differences result mainly from their differing rates of enrichment of silica and iron. The curve for tholeiites is based on lavas of Hawaii and the Galapagos Islands. The calc-alkaline curve is for rocks of the Cascade Range, and the alkaline one for the phonolitic series of Tahiti.

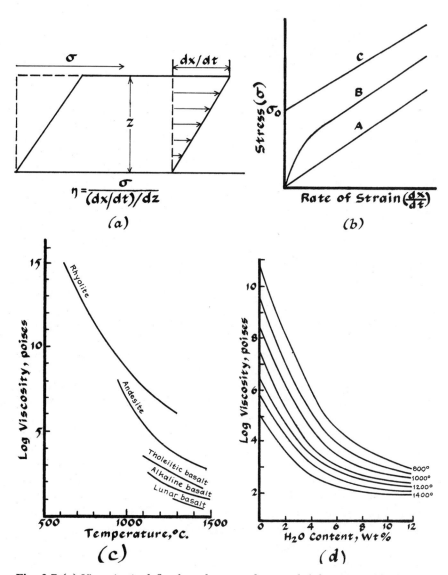

Fig. 2.7 (a) Viscosity is defined as the rate of internal deformation (dx/dt) produced by a shear stress (s) parallel to the base of a layer of thickness (z). This relation **(b)** is linear for most common *Newtonian* substances *(A)*, but many natural silicate melts are *non-Newtonian* in the sense that the relationship is not linear *(B)* or, as in *C*, a minimum stress (s_σ) must be applied before the substance will deform permanently. The viscosities of most magmas decrease with increasing temperature **(c)** and water content **(d)**. The latter is shown for a rhyolitic magma.

The presence of crystals has an added effect. The term *effective viscosity*, η_{eff}, is used for a magma with suspended solids or bubbles of gas. Its value can be estimated from the equation:

$$\eta_{eff} = \eta_o(1 - 1.7\phi)^{-2.5}$$

where η_o is the viscosity of the liquid fraction alone, and ϕ is the volumetric fraction of solids. The exponential term shows that although there is no sharp transition between the solid rock and liquid, effective viscosity increases at an accelerating rate as crystallization advances, especially after solids are so numerous that they are in contact with one another and hinder flow. Beyond this point, known as the *critical melt fraction*, viscosity increases precipitously. Very few magmas reach the surface with less than this critical melt fraction, and those that do are so viscous that they form blocky domes that can scarcely flow, even on steep slopes.

2.7 ▲ Mechanisms of Crystal Fractionation

As we have seen, crystallization can produce significant compositional differences only under conditions that permit crystals to be mechanically separated, or *fractionated*, from the liquid. This is normally achieved by some form of density segregation. For example, ferromagnesian minerals having a density of 2.9 or more may settle out of a liquid with a typical density of 2.7, whereas plagioclase with a density of 2.6 may float. Many lavas contain large crystals that are thought to have been concentrated by a gravitational process of this kind. For example, certain basic volcanic rocks, such as *picrites*, are very rich in olivine phenocrysts, and *ankaramites* contain large numbers of crystals of both olivine and calcium-rich pyroxene. The proportions of olivine or pyroxene in these rocks are far greater than the amounts that could be precipitated from a normal basaltic liquid; the liquids must acquire crystals from some other source. Many of the coarse-grained inclusions in lavas probably result from crystals accumulated in this way.

Gravitational segregation is not the only mechanism of fractionation. The compositions of magmas can also evolve when moving magma flows over static crystals growing on the walls of a magma chamber or volcanic conduit. Regardless of which is moving, crystals or liquid, differentiation normally requires some form of relative motion of the solid and liquid phases. Some of the ways this happens have been deduced from observations in the field, others by modeling crystallization in the laboratory.

2.8 ▲ Modeling Magmatic Processes

Although the Hawaiian lava lakes mentioned earlier have provided an extraordinary opportunity to observe crystallization under natural conditions, events of this kind are rare and illustrate only a limited range of processes. A more convenient way of studying magmas is to simulate natural conditions in a laboratory experiment where crystallization can be controlled and closely observed. This technique permits one to study an essentially limitless range of chemical and mineralogical compositions. Small amounts of material, usually less than a gram, are heated in a furnace in which the temperature, pressure, and oxidation state can be adjusted according to the natural conditions one wishes to reproduce. Modern equipment is now capable of reaching temperatures and pressures equivalent to those at deep levels of the mantle so that one can determine how minerals melt or crystallize from a liquid of a given composition under a wide range of conditions. In this way, much can be deduced about how natural magmas are produced by melting or evolve by crystal fractionation. A serious limitation of this technique is that experiments must be run for relatively brief times and at constant temperatures. Following the course of a differentiating liquid requires that repeated runs be made with appropriate liquid compositions at successively lower temperatures. In such a small sample, the natural mechanisms of crystal fractionation cannot be simulated.

This problem has been surmounted, at least in part, by modeling crystallization processes, either mathematically using the thermodynamic and physical properties of liquids and crystals or by laboratory analogs. One can calculate, by means of a simple mass balance, the proportions of minerals that must be removed from (or added to) a given parental magma to produce a derivative liquid of a particular composition. This is commonly done by using a computer-based calculation that finds the proportions of components that best fit a mass balance of the major elements. In recent years, mathematical methods have been developed to the degree that it is now possible to model crystallization and differentiation of an entire magma chamber. Like all models, of course, these techniques are sensitive to the assumptions built into the calculations. Nevertheless, mathematical simulations and laboratory models have yielded important insights into a variety of magmatic processes.

If a laboratory model is scaled properly in terms of its dimensions and physical properties, such as temperature, viscosity, and density, it can provide a useful analog of the natural system on a greatly reduced scale. Because the scales of size and time of the experiment are many orders of magnitude less than those of nature, the physical properties of the model must be scaled accordingly. This is best accomplished by using liquids with viscosities proportionately less than those of the magma they are designed to represent.

The example shown in Fig. 2.8 was designed to model the behavior of a magma chamber in which the density of the evolving liquid declines with advancing crystallization. Starting with a solution saturated with a simple compound, in this case sodium carbonate, cooling caused the liquid to precipitate crystals of the dissolved salt. As crystals grew on the walls, the adjacent liquid became more dilute and less dense. This buoyant, depleted liquid rose along the walls and collected under the roof, and after a few hours, the liquid had separated into two distinct parts. The upper zone consisted of stable, dilute liquid that increased in temperature and density downward, while the main mass of liquid below maintained a more uniform temperature, composition, and density. A system of this kind may account for the compositional zoning commonly observed when large volumes of magma are discharged in explosive eruptions. The eruption that produced Crater Lake, Oregon, is a spectacular example (Fig. 2.9). A total volume of about 75 km³ was erupted in what seems to have been a single

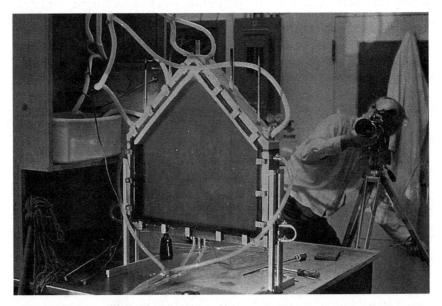

Fig. 2.8 Tanks such as this are used as laboratory analogs to simulate the behavior of magmas in subvolcanic reservoir. The shape is designed to represent an idealized cross section of a particular type of body, and the front and back walls are transparent so that the behavior of the liquid can be observed as it is cooled at the sides, roof, or floor. The magma is represented by an aqueous solution saturated at room temperature with a simple inorganic salt that precipitate crystals on cooling. Crystallization leaves a more dilute liquid of lower density that rises along the walls to accumulate under the roof in the manner postulated for the zoned magma illustrated in Fig. 2.9b.

(a)

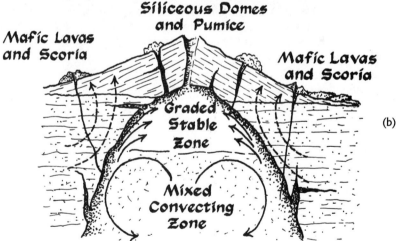

(b)

Fig. 2.9 (a) The main magma discharged during the large eruption that produced the caldera of Crater Lake consisted of white rhyolitic pumice with very few crystals, but near the end of the eruption, it changed abruptly to andesitic scoria containing large amounts of hornblende. The magma had become compositionally zoned **(b)** when crystallization at the walls removed dense components, such as iron, and left a liquid of low density that rose and collected under the roof.

outpouring of rhyolitic pumice with about 72% SiO_2, but an abrupt change occurred in the final stages. The last magma to emerge was a crystal-rich hornblende scoria with an SiO_2 content of only 56.5% (Fig. 2.9a). The main mass of rhyolite must have occupied the upper part of a large body that became zoned by side-wall crystallization and accumulation of a buoyant, differentiated liquid over a denser magma of unknown dimensions (Fig. 2.9b).

In simple experiments of this kind, one can combine theory and observations to gain a better understanding of the types of natural phenomena described in the chapters that follow.

Suggested Reading

Helz, R. T. 1987. Differentiation behavior of Kilauea Iki lava lake, Kilauea Volcano, Hawaii: An overview of past and current work. In B. Mysen, ed., *Magmatic processes: Physicochemical principles*. Princeton: The Geochemical Society, pp. 241–58.
A summary of studies of the most important of the Hawaiian lava lakes. Contains much information on the effects and mechanisms of differentiation.

McBirney, A. R., B. H. Baker, and R. H. Nilson. 1985. Liquid fractionation. Part I: Basic principles and experimental simulations. *Jour Volc Geoth Res* 24:1–24.
Describes modeling of zoned magma chamber, such as that illustrated in Fig. 2.8, by means of laboratory analoges.

Murase, T., and A. R. McBirney. 1973. Properties of some common igneous rocks and their melts at high temperatures. *Geol Soc Amer Bull* 84:3563–92.
A comprehensive summary of the viscosities, densities, and other physical properties of molten igneous rocks.

Ryan, M. P. 1990. *Magma transport and storage*. New York: John Wiley, 440 p.
A comprehensive work on the physical aspects of magmatic processes.

CHAPTER 3

▲

The Volatile Components

In previous chapters, we have discussed the chemical and mineralogical compositions of magmas but have said little about their volatile components. Compared with major elements like silicon, the proportions by weight of magmatic gases exsolved from magmas are relatively small, but their volumetric proportions at low pressures can be enormous. Water and other volatiles play a critical role in governing magmatic behavior, either as dissolved components that affect the viscosity and other physical properties of the liquid or as exsolved gases that determine how the magma erupts. The great quantities of gas released by volcanism can have far-reaching effects, particularly when injected into high levels of the atmosphere where they have a strong influence on much of the earth's climate. Despite their importance, however, the volatile components are one of the least understood aspects of volcanism.

Descriptions of volcanic gases based on visual impressions of untrained observers, particularly in the older literature, are rarely reliable. For example, almost all the references to flames that are found in many popular accounts are erroneous. It is true that some volcanic gases, notably hydrogen, methane, and sulfur, burn when combined with oxygen of the atmosphere, but true flames are very rare in volcanic eruptions. Most of the incandescence observed in high-temperature eruptions is radiation from the molten magma, and clouds described as "smoke" are mostly water vapor with lesser amounts of gas and suspended dust. Sulfur dioxide may give clouds a bluish tint, but in the absence of condensed water vapor, fumes are visually inconspicuous. Because the distinctive odors of sulfur gases are easily detected, even in small concentrations, the importance of these gases is often over estimated. Carbon dioxide or carbon monoxide, on the other hand, are rarely noticed because they are both colorless and odorless.

3.1 ▲ Compositions of Volcanic Gases

The reason why less is known about volcanic gases than about the other components of magmas is that they are much more difficult to collect and study. In only a few cases have samples been obtained directly from erupting vents. Apart from the obvious problems of collecting samples close to their source, it is difficult to be sure that the gases have not been modified by contact with the air. Moreover, the proportions of gaseous species in a sample when it is analyzed in the laboratory are not necessarily those that were present at magmatic temperatures. To interpret the proportions of gases in an analysis, one must know to what degree they have re-equilibrated under conditions of low temperature and pressure.

Gases emitted at high temperatures are of special interest because they tend to be the least contaminated. Samples have been collected from high-temperature sources at Kilauea, Hawaii, as well as at Etna in Sicily and at Ardoukoba and Erta Ale in the Afar Depression of East Africa. These are exceptional, however.

Modern techniques of remote sensing make it possible to measure the proportions of volatile components in fume clouds while observing eruptions from a safe distance. The emission spectrum given off by incandescent vents is used to detect volatile elements at very high temperatures, and measurements of the absorption spectrum of gases provide a way of determining the proportions of the major components in eruption columns.

Excellent analyses can be obtained from bubbles of gases contained in rapidly quenched lavas. Lavas erupted on the sea floor, for example, still contain water and other volatiles in solution or trapped in small bubbles. The total amounts of volatiles can be determined from the loss of weight when the sample is crushed and raised to high temperatures, and the proportions of the various components can be determined by collecting and analyzing the exsolved gases.

Wherever reliable samples are obtained from high-temperature sources, they consist of a combination of a limited number of major components—H, C, O, S, Cl, F, and N (Table 3.1). H_2O normally accounts for 70 to 99 volume percent, but in exceptional cases, especially in basalts, it may make up as little as 50%. Next in order of decreasing abundance come CO_2, SO_2, H_2S, CO, COS, CH_4, HCl, HF, H_2, O_2, S_2, N_2, CS_2, SO_3, NH_4, B, Br, volatile metallic salts, and rare-earth elements. N_2, which may be present in important amounts, is almost entirely of

TABLE 3.1					
Compositions of Some Typical Volcanic Gases					
	1	**2**	**3**	**4**	**5**
CO_2	48.6	40.9	47.0	65.9	7.4
CO	1.4	2.45	2.6	–	1.26
H_2	0.49	0.75	34.1	15.2	0.44
CH_4	–	–	–	15.9	–
HCl	0.04	–	2.9	–	–
Cl_2	–	–	–	–	10.50
F_2	–	–	–	–	4.29
S_2	0.04	–	–	–	2.89
H_2S	–	–	–	1.7	–
SO_2	11.5	4.40	13.4	–	–
SO_3	0.04	–	–	–	–
O_2	–	0.0	0.0	–	–
N_2 + Ar	–	8.30	0.1	0.07	2.44
H_2O	36.9	43.20	–	–	70.81

1. Kilauea, Hawaii, crack in crust on lava lake (Shepherd, 1938), recalculated to exclude atmospheric components by Nordlie (1971).
2. Nyiragongo lava lake, Zaire (Chaigneau, Tazieff, and Fabre, 1960).
3. Surtsey volcano, Iceland (Sigvaldason and Elisson, 1968). The gases shown here made up 13.84% of the total discharge. The other 86.16% was H_2O.
4. Fumarole, Showa-shinzan, Japan (Matsuo, 1961). The gases shown here made up only 1.95% of the total discharge. The other 98.05% was H_2O.
5. Gas extracted from dacitic pumice, Mont Pelée, West Indies (Shepherd, 1938).

atmospheric origin; it is usually accompanied by other gases, such as argon, in the same proportions as these gases have in the atmosphere.

The proportions of volatile components vary widely with temperature and the amount of contamination with atmospheric gases. Most samples are collected at temperatures of a few hundred degrees, and to determine the original proportions of gases at depth, one must allow for the effects of falling temperature and pressure.

At atmospheric pressure and magmatic temperatures, SO_2 is the dominant sulfur compound, but falling temperature favors reaction with H_2O to form H_2S. The effect of decreasing pressure at constant temperatures is in the opposite direction. If these reactions take place in a system that is open to oxidation, the change from SO_2 to H_2S is also one from a reduced assemblage containing CO and H_2 to a more oxidized one with CO_2 and H_2O.

The proportions of gases can change radically during the course of an eruption. Immediately before an eruption, there may be a marked increase in the ratios S/C, SO_2/CO_2, S/Cl, as well as the total amount of HCl. At the same time, the ratio of He to CO_2 decreases. Studies of Mt. Etna showed that the amount of water decreased abruptly during short bursts of hot gases.

3.2 ▲ Sources of Magmatic Gases

The volatiles in magmas have two principal sources (Fig. 3.1). They may be "juvenile" components derived from the mantle sources where magmas are generated, or they may come from shallower, external sources. Much of the H_2O in volcanic gases is meteoric water absorbed as the magma rises through fractured, porous rocks of the crust. It can also be introduced from subducted crustal rocks. Much of the water carried to mantle depths in hydrated rocks and sediments eventually returns to the surface in the eruptive products of subduction-related volcanoes.

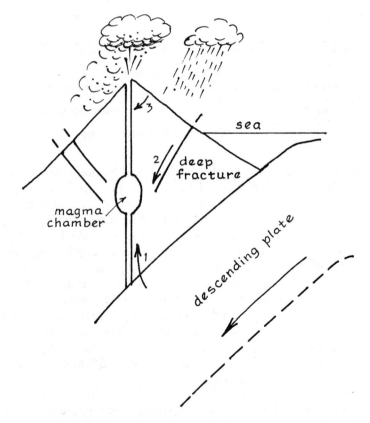

Fig. 3.1 The water in magmas comes from several sources. In the case of subduction-related magmas, a small part comes from subducted igneous rocks *(1)*, and a much greater amount comes from meteoric water infiltrating into the magma through crustal rocks *(2)* and rain or snow *(3)*. The diagram is not to scale.

The isotopic composition of volcanic gases can be used to estimate the relative proportions of gases of different origins. Mantle-derived water has characteristic proportions of the isotopes of hydrogen (^1H and ^2H), and oxygen (^{18}O and ^{16}O). Compared with standard sea water, it has about 8% less deuterium (^2H) and about half a percent more ^{18}O. In contrast to the more or less constant isotopic composition of magmatic water, the proportions of isotopes in meteoric water have a wide range. Groundwater, for example, has less deuterium than sea water, and its oxygen/isotopic ratio varies with latitude and the degree to which the oxygen has been exchanged with the soil and rocks through which it infiltrates.

Magmatic water is a relatively important component of basalts erupted from oceanic ridges and other rifts, whereas water of surface origin is much more prominent in the products of subduction-related volcanoes. In the latter, recycled meteoric water accounts for 90 to 99% of the total, and a large part of the remainder is mantle-derived water contained in subducted lavas and sediments of the oceanic crust.

The isotopic ratios of carbon ($^{12}C/^{13}C$), sulfur ($^{32}S/^{34}S$), nitrogen ($^{14}N/^{15}N$), and the proportions of the rare gases—He, Ne, Ar, Kr, Xe, and Rn—can also be used to identify the relative contributions from different sources. Most of the carbon dioxide in volcanic gases comes from the mantle, but it may be augmented by carbon picked up from carbonate-rich rocks through which the magma rises. The large amounts of carbon dioxide in many Indonesian volcanoes, for example, come from Mesozoic limestones beneath the volcanic arc. In addition, organic matter contributes carbon to meteoric water, which, in turn, is absorbed by magmas at shallow levels of the crust. Carbon from these sources is easily detected because it is much richer in ^{13}C.

The proportions of magmatic and meteoric water vary with the stage of activity of individual volcanoes. A marked increase in the proportions of mantle components is often a sign that new magma has risen into a dormant volcano and that an eruption may be expected. As the proportion of magmatic components increases, the gases tend to be less oxidized, so ratios, such as CO/CO_2, H_2/H_2O, and H_2S/SO_2, increase as well.

3.3 ▲ Volumes and Rates of Emission

The amount of gas produced by individual volcanoes is largely a function of the nature of the magma and the type of activity. Few basalts contain more than 1% dissolved water, whereas lavas of intermediate compositions, such as andesites, may contain up to 6%, and some rhyolites have as much as 7% total volatiles. Alkaline magmas tend to have larger proportion of carbon dioxide, chlorine, and fluorine than do tholeiitic or calc-alkaline magmas. These are only generalizations, however. The compositions and amounts of volatiles vary widely with time and from one volcano to another.

The total amount of carbon gases released each year by volcanic eruptions on land is estimated to be about 31 million tons of CO_2 plus another 34 million tons released by diffusion through the crust. The total for SO_2 is between 0.5 and 1 million tons. The amount of carbon dioxide and sulfur added to the atmosphere by large volcanic eruptions are of about the same magnitude as those contributed by automobiles and thermal power plants. An individual volcano can continue to emit fumes for decades at rates that exceed those of all the coal-burning power stations of a large metropolitan area.

The quantities of water vapor and other volatiles released in the summit regions of volcanoes over prolonged times are usually many times greater than those that can be explained by magma reaching the surface during a comparable period. It is common for volcanic vents to expel conspicuous fume clouds for years or even decades without erupting fresh magma. For example, each day the summit crater of Merapi on the island of Java gives off approximately 3,000 tons of CO_2, 400 tons of SO_2, 250 tons of HCl, and 50 tons of HF. Measured emission rates of SO_2 as high as 420 tons per day have been estimated for some of the Central American volcanoes. It is calculated that the SO_2 emitted from Pacaya volcano in Guatemala during a period of activity in 1972 (270 tons per day) was about five times that which could be supplied by degassing of the amount of lava erupted during the same period. It is thought that these large amounts of gas are supplied by a source in which new magma is continually circulating through the reservoir.

When gases such as these are emitted under a lake in the summit crater of a volcano, the water acts as a large condenser that absorbs heat and volatile emanations given off by vents on the crater floor. If the input of heat exceeds a certain limit, the lake may become unstable and erupt a combination of gas, water, and mud (see Chapter 9). The compositions of lake waters differ widely depending on the components contributed by volcanic emanations and those leached from rocks exposed to the hot acidic water.

In addition to the conspicuous gases released from active vents, large amounts are given off from more dispersed sources. Gases emitted by small fumaroles or diffusing through the porous soil on the flanks of volcanoes are easily overlooked or mistakenly dismissed as trivial. Although the total amounts given off in these sources are difficult to quantify, they are probably of the same magnitude as those released from craters. The summit plume of Etna produces about 35,000 tons of CO_2 each day, but this value would be doubled if one could take into account the diffuse discharge of gas from the flanks of the edifice. Most of the large amounts of gas, particularly CO_2 seeping from the ground, are unnoticed, but on occasion they become very evident. At Mammoth Mountain in California, for example, it was recently observed that hundreds of trees were dying, and on investigation, it was found that they were being killed by gases diffusing through the soil, presumably from magma under the nearby caldera of Long Valley. Measurements showed that in one area CO_2 was being given off at a rate of 1,100 tons per day, and in places the air was 60% CO_2.

3.4 ▲ Solubility Relations

Three important factors govern the behavior of dissolved volatiles: their initial concentration in the magma, their solubility limit at the prevailing temperatures and pressures, and the manner in which a separate gas phase is formed. The greater the amount of volatile components in solution, the greater the ability of a magma to expand when it exsolves. This is why water plays a major role, whereas carbon dioxide, which is much less soluble, is normally of less importance.

The amount of water vapor in solution also varies according to the proportions and compositions of any crystals the magma may contain. As magmas stored at intermediate levels adjust to the temperatures and pressures at crustal conditions they begin to crystallize anhydrous minerals, such as olivine, pyroxene, and plagioclase, and the dissolved volatiles are concentrated in diminishing amounts of liquid. Their concentration can be expressed in terms of the following simple equation:

$$X_r (1 - f) = X_i \qquad (1)$$

where X_i is the initial concentration and X_r is the concentration after a fraction of crystals, f, have been removed ($0 \le f \le 1$).

On the other hand, the concentration of water in the remaining melt fraction may not increase or may even decrease if large proportions of hydrous minerals, such as amphibole and mica, crystallize and take water into their crystalline structure. The amphiboles, most of which are stable up to pressures of 20 to 25 kbar and temperatures of around 1000°C, contain about 2% water. The concentration of water in a magma crystallizing a mineral like amphibole is expressed by the following equation:

$$X_r (1 - f) = X_i - X_{OH} f' \qquad (2)$$

where X_{OH} is the water content of the hydrous minerals, f' is the proportion of hydrous minerals, f is the total fraction of crystals (both hydrous and anhydrous), and $0 \leq f' \leq f \leq 1$.

If, at some later stage, these conditions change, the hydrous minerals may become unstable and break down. This can happen, for example, when pressure drops during ascent of the magma or when an influx of new magma causes an abrupt increase of temperature. Hydrous minerals may then be destabilized and liberate the water they held in their structure. In the theoretical example shown in Fig. 3.2, a magma that is assumed to have crystallized 30% anhydrous crystals and 10% amphiboles can yield 0.33% water if the amphibole suddenly breaks down to pyroxene. This amount of water is in addition to that already contained in the liquid fraction, 1% in this example. Experimental studies show that, in the case of amphibole, this could happen at depths as great as 5 to 8 km. Micas, which take about 4% water into their structure, could play a similar role, even though they normally crystallize at somewhat lower temperatures. The water pressure generated in this way may be sufficient to disrupt material plugging a volcanic vent.

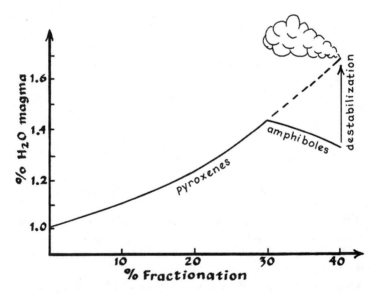

Fig. 3.2 Crystallization of anhydrous minerals, such as pyroxene, results in an increase in the amount of water concentrated in the remaining liquid, but the appearance of amphibole has the opposite effect. If the amphibole breaks down to pyroxene, however, it liberates water and can destabilize the magma. (From Bardintzeff, J. M., and B. Bonin. 1987. *J Volc Geoth Res* 33:255–62.)

3.5 ▲ Effects on Physical Properties

We noted earlier that volatiles have a marked effect on the viscosity of magmas (Fig. 2.6e). Density is also affected, but to a lesser degree. As a magma crystallizes and the dissolved volatile components are concentrated in a diminishing volume of melt, its density and viscosity can change accordingly. As shown in Fig. 3.3, these effects can be calculated for a given magma to show how the concentrations of water in

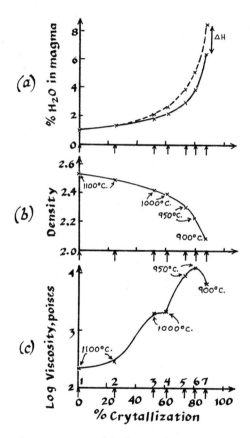

Fig. 3.3 Evolution of water content **(a)**, density **(b)**, and viscosity **(c)** of a magma evolving by crystal fractionation. Calculations are based on a series of samples, CP_1 through CP_7, from the Chaine de Puys of central France. In **(a)**, the solid curve depicts the increasing water content caused by crytallization of all the observed anhydrous and hydrous minerals. The dashed line indicates the effect when amphibole breaks down to pyroxene and an amount of water, ΔH. (From Bardintzeff, J. M., and B. Bonin. 1987. *J Volc Geoth Res* 33:255–62.)

an evolving series of magmas influence their eruptive behavior. We shall see other examples in the chapters to come when we examine the eruptive processes for lavas and pyroclastic rocks.

Suggested Reading

Allard, P. 1983. The origin of hydrogen, carbon, sulphur, nitrogen and rare gases in volcanic exhalations: Evidence from isotope geochemistry. In H. Tazieff and J. C. Sabroux, eds., *Forecasting volcanic events*. Amsterdam: Elsevier, pp. 337–86.
An excellent review of volcanic gases and their use in forecasting eruptions.

Burnham, C. W. 1979. The importance of volatile constituents. In H. S. Yoder, Jr., ed., *The evolution of igneous rocks: Fiftieth anniversary perspective*. Princeton, NJ: Princeton University Press, pp. 439–82.
A broad overview of the important role of volatiles in magmatic processes.

Giggenbach, W. F. 1996. Chemical composition of volcanic gases. In R. Scarpa and R. I. Tilling, eds., *Monitoring and mitigation of volcanic hazards*. Berlin: Springer-Verlag, pp. 221–56.
An excellent summary of the compositional relations of volcanic gases.

Jaggar, T. A. 1940. Magmatic gases. *Am J Sci* 238:313–53.
A classic early work on the gases in Hawaiian lavas.

CHAPTER 4

Structural Controls and Triggering of Eruptions

▼

Once a batch of magma has risen into the lithosphere, it is staged to continue its ascent and reach the surface in a volcanic eruption. Not all magmas erupt, however; many intrude the crust without ever reaching the surface. What special combination of conditions determines whether magmas pass this "point of no return?"

4.1 ▲ Tectonic and Structural Controls

As we have seen, the great majority of active volcanoes are situated in tectonic regimes in which the conditions in the lithosphere are governed by a distinct structural regime. Where the stresses are extensional they facilitate the ascent of magma; where compressional, they impede it. However, even under the extensional condition at oceanic ridges and continental rifts, only part of the magma erupts at the surface; a large fraction, probably more than half, is intruded as dikes and sills. At convergent plate boundaries, volcanism is confined to well-defined zones where upwelling mantle and flexure cause the lithosphere to be in tension (Fig. 4.1), and in many cases, volcanism is much less important than plutonic intrusions. The long episode of Circum-Pacific plutonism lasting through most of the Mesozoic and early Cenozoic eras was one of the great magmatic events of the earth's history, and yet the amount of volcanism during that period was relatively minor. Why some periods are dominated by plutonic intrusions and others by volcanism has never been fully explained, but it must be related in some way to global tectonic conditions.

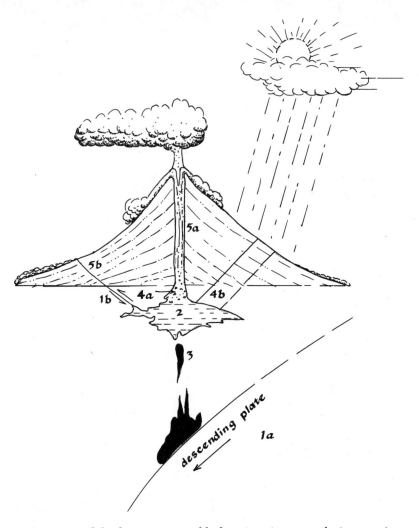

Fig. 4.1 Some of the factors responsible for triggering an explosive eruption can be illustrated using subduction-related volcanoes as an example. Tectonic factors *(1)* include the rate of subduction *(1a)* and formation of dilational fractures that facilitate rise of magma through the overlying lithosphere *(1b)*. Processes in the magma chamber include differentiation and accumulation of volatile-rich magma in the upper levels of stratified magma chambers *(2)* and influxes of new magma *(3)*. Volatiles contributing to pressure in the magma chamber include water of magmatic *(4a)* and meteoric *(4b)* origin. Additional factors are the resistance of solidified lava in the conduit to the rise of new magma *(5a)* or the strength of the overlying structure forcing magma to break out on the flanks *(5b)*. The diagram is schematic and not to scale; the thickness of the overlying structure is probably hundreds of meters, and the depth of the magma chamber is thought to be a few kilometers but may be as much as 30 km. Dehydration of the oceanic crust takes place at 100 km or more. (From Bardintzeff, J. M. 1986. *Volcans et magmas*, Le Rocher, Paris, Monaco.)

On a more local scale, the locations of individual vents within a volcanic field or province are governed by shallow structural features of the crust. The spatial relations to faulting are most clearly seen in regions where numerous monogenetic cinder cones are linearly distributed along extensional faults or rifts. Examples include the volcanoes of Iceland, Hawaii, or the Basin and Range Province of the United States. Large transverse faults provide a less favorable setting, except where they abruptly change direction or intersect other active faults. Basaltic cones are also common in wide zones of transcurrent movement where extensional faults are oriented obliquely to the main sense of shear. Thus, the general rule is that volcanism tends to be concentrated along axes parallel to the direction of the maximum horizontal stress and perpendicular to the direction of least stress in the lithosphere.

4.2 ▲ Triggering by Tectonic Events

If magma is ready to be tapped in a high-level reservoir, even a small change of regional stress, usually associated with an earthquake, can disturb the stability of the system and bring about an eruption. The eruption of Mt. St. Helens in 1980 is a good example. It was preceded by two months of increasing seismicity, including an earthquake of magnitude 4.2 that occurred 4 km below the volcano on the 20th of March. These early earthquakes seemed to be caused by tectonic forces, for they were distributed along a northwest-trending line that extended well beyond the immediate vicinity of the volcano. A series of smaller shocks followed, but unlike the aftershocks that normally follow tectonic earthquakes, they did not decline but increased in frequency to more than 30 per hour. By the time the first ash eruption began on March 27th, they were almost continuous. At the same time, their focus rose to shallower depths, indicating that they were related to a rise of magma that was set in motion by the tectonic earthquakes. Many small shocks followed as the magma rose into higher levels of the volcanic system.

Similar conditions are occasionally seen in Hawaii where earthquakes begin at depths of about 50 km below the volcanoes and over months or years become shallower and more numerous. Immediately before the outbreak of an eruption, the shocks are almost continuous and take on a distinctive character, known as *harmonic tremor*, that is associated with magma flowing through channelways within the volcano. The tremor may continue during the following stages of eruptions, particularly when very fluid lava is discharged from fissures.

The nearly simultaneous eruptions of two or more volcanoes, such as the Soufrière of Saint Vincent (May 7, 1902) and Mount Pelée (May

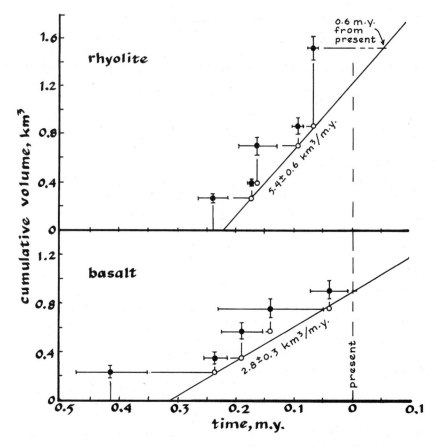

Fig. 4.2 Cumulative erupted volume versus time for rhyolites and basalts in the Coso Volcanic Field. Error bars on dots show estimated precision of ages and cumulative volumes of eruptive episodes. Circles have been plotted by projecting cumulative volumes to times of succeeding eruptions. A basaltic eruption is due soon, but a rhyolitic one is for the future. (After Bacon, C. R. 1982. *Geology* 10:65–9.)

8, 1902) or Unzen in Japan and Pinatubo in the Philippines, both of which erupted in June of 1991, may result from a regional tectonic "crisis" or from stresses in the earth's crust, such as earth tides, that fluctuate with time. Many observers have commented on the tendency for eruptions to be initiated or become stronger at times of full moon when the tidal stresses in the crust are greatest.

In certain regions, such as the Great Basin of the western United States, the timing of eruptions seems to be directly related to the rate of crustal extension. This is best seen in the Coso Volcanic Field of southern California where the repose interval between eruptions has been found to vary with the volume of the preceding eruption. Although both basalt and rhyolite are erupted more or less contemporaneously, the rhyolite appears as small domes near the center of the field, whereas basalt is largely restricted to scoria cones around the margins. Each type of magma has its own time–volume relation (Fig. 4.2). It is postulated that the magmas coexist in a zoned magma chamber at a depth of 8 km or more and that the overlying crust is deforming by tension at a constant rate. Because large eruptions relieve the dilational stress in the crust more than small ones, the time that must elapse before a critical stress is reached is a linear function of the volume of magma evacuated.

Although the frequency of tectonic events may be nearly constant on the scale of tens or hundreds of thousands of years, this is not necessarily true for longer periods. Using a record of activity spanning a few centuries, one can extrapolate volcanism on the assumption that rates and directions of tectonic forces will remain nearly constant, but this assumption may not be valid on a longer scale. Like seismicity, volcanism has a temporal distribution that is strongly episodic.

4.3 ▲ Opening of Vents

The forces by which new eruptions are opened may be exerted either by the magma itself, by increased gas pressure, or by some combination of the two. In the case of basalts, the volatile content of the magma is normally less than that of more siliceous magmas, and the pressure of volatiles plays a subordinate role in overcoming the strength of overlying rocks. This is why the opening eruptions of basaltic magma tend to be only mildly explosive. When the Mexican volcano Paricutin was born in 1943, the first eruption began with a quiet rise of basalt, and although scoria was ejected, it was not associated with violent explosions. In a similar way, when eruptions begin on the rift zones of Kilauea, they open with a quiet upwelling of lava that gradually develops into spectacular fountaining as the vent becomes a widening fissure and the rate of discharge increases.

In contrast, the more volatile-rich, differentiated magmas, such as dacites and rhyolites, commonly break out with powerful explosions of gas and steam, a large part of which is heated groundwater. The first outbursts are usually heavily laden with lithic debris eroded from the conduit walls by the high-velocity flow of gas. If the gas is distributed throughout a large mass of magma, the violence of the eruption increases in magnitude with time; but more often, most of the gas is concentrated in the upper part of the column, and the opening explosive phase may quickly give way to quiet extrusions of viscous lava or domes.

The transition from explosive to effusive discharge may be the result of concentration of volatiles in the upper levels of the magma reservoir, but it can also be due to a decrease in the eruption rate as the overpressure on the magma declines. Decreasing the rate at which magmas ascend allows gases to exsolve and escape without developing internal gas pressures large enough to cause fragmentation.

4.4 ▲ Triggering by New Injections of Magma

The long-term activity of most volcanoes is governed by the rate of supply of magma, and the spacing of eruptions is directly related to the volume of magma coming from the mantle. When eruptions are frequent, they tend to produce smaller volumes of lava than if they follow a long period of repose (Fig. 4.3). These relations suggest that magma is supplied at a nearly constant rate and that an eruption occurs when the volume of magma passes some critical limit governed by the structure of the volcano. Although the long-term rate of production of magma may be nearly constant, the intervals of time between eruptions differ widely, even in the same volcano. The timing depends on the physical nature of the magma reservoir and on how long it can store magma before the surrounding rocks yield to magmatic stresses that propagate fractures to the surface.

The main source of these stresses is probably injections of new magma that increase pressures within the reservoir. If the volcano is large and the magma rises high into the conduit, the entire volcanic structure swells, and when the strength of the rocks is exceeded, fractures are opened. This phenomenon is seen quite clearly in Hawaii where precise geodetic measurements show that the upper slopes of Kilauea bulge and tilt outward as the magma flows into a network of dikes and sills a kilometer or two beneath the summit. When the stresses reach some critical value, fissures open, lava pours out on the surface, and the volcano subsides. The surface does not always return to its original level, however; some of the magma usually remains in the subsurface adding a small increment to the elevation of the surface, so that with time, the volcano grows by a combination of surface flows and inflation.

A similar mechanism is responsible for triggering eruptions from large stratocones. As a volcano grows to higher and higher elevations, magma standing in the central conduit exerts increasingly greater stresses on its walls. Beyond a certain limiting height, less energy is required for the magma to propagate radial fractures and break out on the flanks of the cone than to reach the summit crater. This is why

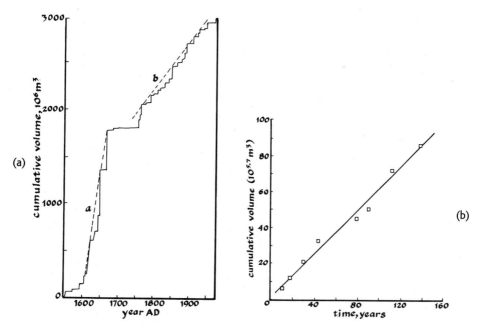

Fig. 4.3 (a) Cumulative output of magma from Mt. Etna by flank activity since 1535. In the period represented by section **a** the average output rate was 0.83 m³ sec⁻¹, but around the beginning of the 18th century, this changed to 0.17 m³ sec⁻¹ as seen in section b. **(b)** Cumulative output of pyroclastic ejecta from Mt. Etna since 1868. Note that these are volume–predictive relations because the volume of the next eruption can be predicted from the time elapsed since the last one. This is in contrast to eruptions triggered by tectonic strain (Fig. 4.2), where the time of the next eruption can be predicted from the volume of the previous one. (**a**, After Wadge, G., G. P. L. Walker, and J. E. Guest. 1975. *Nature* 255:385–87, **b**, After Palumbo, A. 1998. *J Volc Geoth Res* 83:167–71.)

flank eruptions become increasingly common as large, mature volcanoes become higher.

One should distinguish the case of a simple plugged conduit above a shallow body of magma from that of a large volcanic edifice forming the roof of a deeper magma reservoir at a depth of more than a kilometer or so. In the latter case, a much greater force must be exerted by the magma to break through to the surface. Failure of the roof is more likely to result from gravitational forces causing the roof to settle into a reclining reservoir of magma. When this happens, the magma usually rises through fissures created by passive subsidence of the roof (Fig. 4.4).

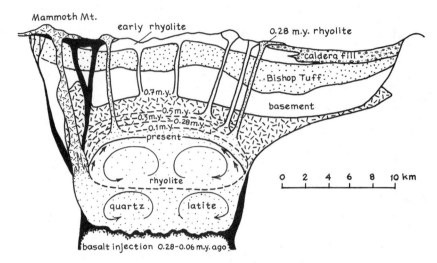

Fig. 4.4 The caldera of Long Valley is an example of subsidence of the roof of a large reservoir. Curved dashed lines show the depths of crystallization at different times (millions of years ago). The upper levels of the magma were rich in silica and volatiles and less dense than the overlying crustal rocks. Pressure exerted on the roof by the gas-rich magma resulted in doming and eventual failure along ring faults. (After Bailey, R. A., G. E. Dalrymple, and M. A. Lanphere. 1976. *J Geoph Res* 81:725–44.)

Injections of new magma can also trigger eruptions by upsetting the thermal, chemical, or mechanical equilibrium of older magma in a shallow reservoir. New magmas coming from deeper, hotter sources can suddenly raise the temperature of the cooler resident magma causing it to convect and vesiculate. Certain petrographic features, such as resorbed or reversely zoned phenocrysts, suggest that events of this kind are quite common, especially in explosive eruptions.

A question of practical importance to forecasting is how much time elapses between such an injection of new magma and an ensuing eruption. In many volcanoes, such as those of Hawaii, this time interval may be measured in weeks or months, but in others, it seems to be much shorter, possibly days or hours. Evidence for such short delays is found in the phenocrysts and inhomogeneities preserved in the magma. For example, the felsic magma erupted in 1645 BC from the Aegean volcano Santorini contained clots of mafic magma several millimeters in size. If the time interval had been greater than a few weeks, these would have crystallized or been absorbed. Similar conclusions have been drawn for the Soufrière volcano of Guadaloupe. A maximum of six days would be necessary to form the rims that grew around phenocrysts if one assumes a reasonable growth rate of about 10^{-8} cm sec^{-1}. The sharp contacts between closely intermingled andesitic and rhyolitic glass indicate that the two magmas could not have been in contact for more than about 10 hours if the rate of diffusion had a normal value of about 10^{-12} cm^2 sec^{-1}. Similar evidence of brief periods following new injections have been found in magmas of the Medicine Lake Highlands of northern California. In all these instances, the time intervals were so short that it would have been difficult to detect the triggering event in time to forecast the resulting eruption.

4.5 ▲ Triggering by Slow Magmatic Differentiation

As activity declines and injections of new magma become less frequent, eruptions are more likely to result from changing conditions in a slowly cooling reservoir of magma. As we observed in Chapter 2, crystallization and differentiation causes the composition and physical properties of the remaining magma to evolve, and, depending on how these changes affect viscosity, density, and volatile content, they may or may not cause the magma to erupt.

The viscosities of magmas normally increase with declining temperature, partly in response to increasing proportions of silica but even more through the effect of suspended crystals. When the fraction of solids exceeds a critical limit of about 50%, magma is so viscous that it is unable to force its way to the surface. The ability of magma to rise through the crust is also a function of its density. The densities of most magmas decline with differentiation, mainly as a result of diminished proportions of Fe and Ti and increasing concentrations of volatile components. A magma intruded at a level where it is in gravitational equilibrium with the rocks of the crust may later become unstable as it evolves to compositions of lower density. Basic tholeiitic magmas are the main exception to this rule; an enrichment of iron in the early and middle stages of differentiation causes their densities to increase (Fig. 4.5), and it is thought that this accounts for the relative scarcity of lavas of intermediate compositions in volcanoes of this type. The effect of water on density, however, can outweigh that of other components, especially at shallow depths where volatiles begin to vesiculate.

4.6 ▲ The Role of Volatile Components

A strong correlation has been found between the magnitude of eruptions and the length of the preceding interval of repose. Almost all very

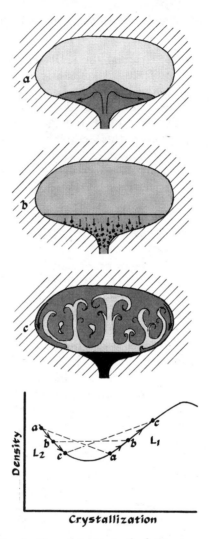

Fig. 4.5 Because of the density minimum in the basic range of tholeiitic magmas (Fig. 2.6a), it is possible that a new injection of hot, basic magma, L_2, could be denser than a slightly more evolved magma, L_1, that has already begun to differentiate in a shallow chamber. On intrusion, the dense new magma would pond at the base of the reservoir **(a)**, but on cooling and crystallizing, the new magma could fractionate olivine and become lighter while the original magma could fractionate olivine and plagioclase and evolve to a more-iron-rich and denser composition **(b)**. If this process continues, the density relations might be reversed **(c)**, in which case the two liquids could overturn, and if mixed turbulently, the resulting magma could have an intermediate composition and density. The schematic diagram **(d)** shows the density relations with tie-lines connecting the two magmas at each of the stages shown. The resulting pressure changes might cause vesiculation and trigger an eruption. (Based on a model by Sparks, R. S. J., P. Meyer, and H. Sigurdson. 1980. *Earth & Planet Sci Ltrs* 46:419–30.)

large, historic eruptions have come from volcanoes that have been dormant for centuries. As already noted in Chapters 2 and 3, the concentration of volatile components that builds up as magmas slowly cool and crystallize is one of the most important factors governing the initiation of explosive eruptions.

The pressures due to the load of overlying rocks at the level of shallow magma chambers range from about 50 to more than 200 MPa (500 bars to 2 kbars). To this must be added the strength of rocks, which is estimated to be about 7 to 10 MPa for basaltic volcanoes and 12 to 16 MPa for granitic wall rocks. Only when the pressure of magma exceeds these values can it break through to the surface.

One can estimate the water pressure in magma chambers by means of the following equation:

$$aH_2O = kX^2$$

where aH_2O is the activity of pure water, X is the weight percent of water, and k is a constant. In the particular case of a magma of silica-rich composition, the following applies:

$$P = 51.02X^2$$

where P is the pressure in bars. Thus, pressures of 200, 450, and 800 bars correspond to water contents of about 2, 3, and 4%. Depending on depth, pressures in this range could exceed that of the overlying rocks and the strength of lava blocking a vent.

4.7 ▲ Dormancy or Extinction?

How long must a volcano remain inactive before it can no longer return to life? The answer to this question depends on the mechanisms governing the spacing of eruptions in volcanoes of different tectonic settings. Each of the principal factors that determine how eruptions are initiated—tectonic forces, rise of new magma, and increasing pressure of gases—operates on a different time scale and within wide limits that are difficult to evaluate. Tectonic forces external to the magma operate over periods measured in centuries or thousands of years depending on the effects of global processes on the region in which a particular volcano is situated. In a similar way, the rate of supply of new magma governs the long-term rates of volcanism, but, as we have seen, the intervals between eruptions are determined by structural conditions within the volcano and are difficult to quantify. Increases of gas pressure and other processes related to slow cooling and crystallization are even harder to evaluate because we have no way of observing them.

If the record of past eruptions were better, it might provide an empirical way of determining the maximum repose times of particular types of volcanoes. From this information, one could arrive at a statistical basis for saying that any volcano that has not erupted for a time longer than such a period could be considered extinct. Unfortunately, the records of activity of long-dormant volcanoes are not adequate to do this. As we shall see in Chapter 14, this problem poses a serious dilemma for volcanologists attempting to assess volcanic hazards.

Suggested Reading

Crisp, J. A. 1984 Rates of magma emplacement and volcanic output. *J Volc Geoth Res* 20:177–211.
A valuable quantitative study of the rates of magmatism on a global scale.

Simkin, T., and L. Siebert. 1984. Explosive eruptions in space and time: Durations, intervals, and a comparison of the world's active volcanic belts. In *Explosive volcanism: Inception, evolution, and hazards.* Washington, DC: National Academic Press, pp. 110–21.
A statistical analysis of the timing and explosivity of volcanic eruptions.

Smith, R. L., and R. G. Leudke. 1984. Potentially active volcanic lineaments and loci in western conterminous United States. *Explosive volcanism: Inception, evolution, and hazards.* Washington, DC: National Academic Press, pp. 47–66.
An exceptionally thorough study of the relationship between tectonic forces and volcanism in the western United States.

Wadge, G. 1984. Comparison of volcanic production rates and subduction rates in the Lesser Antilles and Central America. *Geology* 12:555–58.
A quantitative study of the relation of volcanism rates of subduction.

Lavas

▼

5.1 ▲ Properties of Lavas

The temperatures at which magmas erupt are closely linked to their composition. They range from a maximum of about 1200°C for basalts to a minimum of 850 to 900°C for rhyolites. These temperatures have been measured in several ways. Direct measurements using a thermocouple thrust into the molten lava, although usually reliable, are dangerous and rarely feasible. Optical measurements based on the infrared radiation emitted by incandescent lava give equally reliable results, provided, of course, they are made on openings through the cooler crust. Even after lavas have cooled and solidified, one can calculate their eruption temperatures from the compositions of minerals and the liquid or glass in which they were growing. Although each of these methods has a degree of uncertainty, the results are broadly consistent.

As noted earlier, nearly all lavas are at least 50% liquid when they appear at the surface. It is rare to find lavas with less than this critical fraction of melt because with any smaller amount magmas become too viscous to flow under normal stresses.

Although silicate melts make up the majority of the lavas discussed here, we should note a few rare exceptions. Lavas and ash of sodium carbonate have been erupted from the volcano Oldoinyo Lengai of East Africa, and flows consisting almost entirely of magnetite are found in the Chilean Andes. Molten sulfur has been erupted from a number of volcanoes, such as Lastarria in Chile and possibly from the active volcanoes on Io, a moon of Jupiter (see Chapter 12). For practical purposes, however, we can ignore these uncommon compositions and confine our attention solely to the silicate magmas that account for the vast bulk of terrestrial lavas.

The form and behavior of lava flows are mainly a function of their viscosity, which, in turn, is governed by composition, temperature, volatile content, and crystallinity, most of which change as the lava moves away from its source and cools. As we saw in Chapter 2, viscosity by definition is the constant of proportionality between the shear stress and flow velocity (Fig. 2.6). Fluids that have this linear relation are said to be *Newtonian*. As they cool, however, most lavas take on a non-Newtonian behavior; that is, their velocity of flow is not directly proportional to shear stress. Some fluids move only when the stress exceeds some critical value known as *yield strength*. A lava that is very fluid and flows rapidly when it first appears becomes more viscous as it cools, and at the same time its yield strength increases, so it eventually comes to rest, even on steep slopes. The rates of increase of viscosity and yield strength on cooling are mainly a function of composition and proportions of crystals. They increase more rapidly in mafic, basaltic lavas than in magmas of more evolved compositions.

The *apparent viscosity* of lava, η_a, can be calculated using the Jeffreys equation:

$$\eta_a = \frac{g\rho a d^2}{3u}$$

where g is gravity, ρ is density, a is the slope of the ground, d is the mean depth between the levees, and u is the mean velocity. The equation provides a reasonably accurate description of the behavior of high-temperature lavas with few crystals but is less useful for lavas that have begun to cool and crystallize. A closer approximation can be obtained from the relation:

$$\eta = \eta_a\left(1 - \left(\frac{W}{2W_b}\right)^{-1/2}\right)$$

where W is the width of the flow and W_b is the width of the levee. Even with this modification, however, viscosities calculated from field measurements are at best approximations because they are a composite value for the entire mass, including both the solid crust and fluid interior.

More direct estimates of the viscosity of fluid magma have been made by thrusting a steel rod through the crust and into the hot interior. The rate at which the rod penetrates under a given force can be calibrated against substances of known viscosity, such as asphalt. The technique yields an apparent viscosity and gives no measure of yield strength, but the latter property can be estimated from the heights of flow fronts. Because a lava can flow only when its thickness is great enough for the gravitational force to exceed its yield strength, S_y, the latter, is related to the thickness, d_c, of a lava at a point where it has ceased to flow. This relation is expressed by the equation:

$$S_y = \frac{d_c}{g\rho a}$$

5.2 ▲ Flow of Lava

Fluid lavas move primarily by internal shear. Near the base, the velocity increases upward until it reaches a constant value (Fig. 5.1a). More viscous lavas advance with a rolling motion like that of a caterpillar tread (Fig. 5.1b). The upper surface moves more rapidly than the

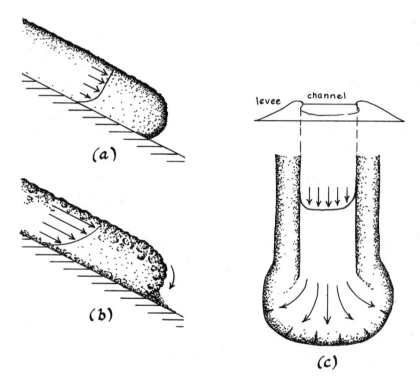

Fig. 5.1 Diagram showing schematic longitudinal and cross sections through a idealized lava flow. Depending on its viscosity and slope angle, the flow may move with a sliding motion in which most of the shear is concentrated near the base **(a)** or a rolling motion in which the upper crust is constantly being overrun like a tractor tread **(b)**. Most viscous flows move between levees and over a basal layer. On reaching the base of the slope, the flow front spreads laterally to form a broad tongue.

interior, so the upper crust is constantly falling down the steep front
of the flow and being overrun as the flow advances. In a similar way,
the faster-moving central channel flows between marginal levees of
blocky scoria.

Relatively hot, fluid lavas flow readily on gentle slopes following
valleys and always seeking topographic lows. As the photographs in
the color insert show, lavas with low viscosities move quickly, sometimes at speeds of tens or, in exceptional cases, up to 75 kilometers
per hour for distances of 50 km or more. They spread to widths of
kilometers and when ponded may reach thicknesses of tens or hundreds of meters. Some of the basalts of Iceland have flowed nearly
100 km, and the Miocene flood basalts of the Columbia Plateau traveled as far as 160 km, covering areas of more than 15,000 km². Because they spread so readily, lavas of these types form broad, gently
sloping *shields* like those of the Hawaiian and Icelandic volcanoes
(see Chapter 10).

The distances lavas flow are primarily functions of their composition, temperature, and rate of discharge. In turn, the rate of discharge
affects the cooling rate and viscosity. Small flows that emerge slowly
lose their heat quickly, become increasingly viscous, and soon come to
rest. The greater supply of heat in rapidly discharged lavas reduces the
rate of cooling and keeps them fluid long enough to flow for greater
distances. Because few lavas are simple Newtonian fluids, they do not
continue to flow indefinitely down topographic gradients. Even
though its interior may still be relatively fluid, a flow can come to rest
on slopes of 20 degrees or more. This is best seen in subaqueous lavas.
A rapid loss of heat from the surface thickens the crust and retards the
advancing flow, but the still-fluid interior can move through tubes that
are insulated from this effect. Where these tubes break through the
crust they form bulbous protrusions or "toes."

5.3 ▲ Modeling the Flow of Lavas

For simple conditions, it is possible to model the behavior of lavas by
computer simulations and calculate the rates at which they would advance and the area over which they will spread. This has been done
with remarkable success in Japan and Reunion. A comparison between the calculated and observed forms of a lava flow, such as the one
in Fig. 5.2, shows how this method can be of great value in assessing
the hazards of future volcanic events.

The flow of lava has also been modeled in the laboratory using material with properties similar to those of natural magmas. If properly
scaled, models of this kind can be used to examine under controlled

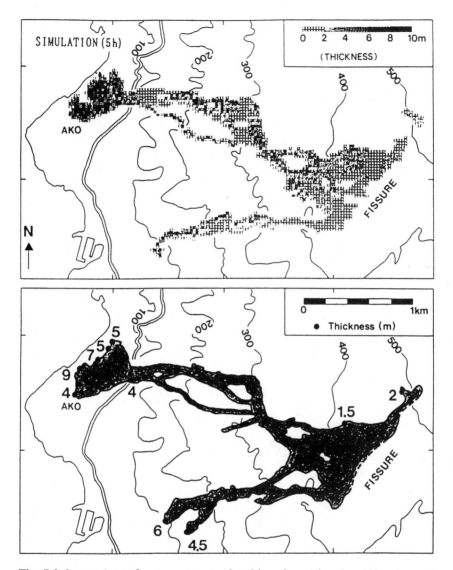

Fig. 5.2 Comparison of a computer-simulated lava flow *(above)* and the observed flow of October 1983 from the volcano Miyakejima, Japan. The simulated flow is shown 5 hours after the beginning of its emission. The different patterns represent calculated thicknesses. The measured thicknesses of the observed flow is shown by numbers near the edges. (From Ishihara, K., M. Iguchi, and K. Kamo. 1990. Numerical simulation of lava flows on some volcanoes in Japan. In J. Fink, ed., *Lava flows and domes*, Berlin: Springer-Verlag, pp. 174–207.)

conditions the individual effects of the numerous factors that govern the flow of natural magmas.

5.4 ▲ Volumes and Rates of Discharge

In a general sense, the rates of discharge of lavas vary inversely with viscosity. Many basaltic fissure eruptions discharge hundreds or thousands of cubic meters per second, whereas the lavas of viscous domes emerge at rates of only a few tens of cubic meters per day. The Roza member of Columbia River flood basalts, for example, is thought to have discharged a total volume of about 1,300 km^3 covering an area of 40,300 km^2 in 10 years or about a third of a cubic kilometer per day. The only comparable eruption for which we have measured values is that of the Icelandic volcano Laki, which, though smaller, poured out an average of 5,000 m^3 per second or about 0.4 km^3 per day during the first days of the great eruption of 1783.

Eruptions of lava can vary in intensity over hours, days, months, or even years. The rate of emission usually increases rapidly at first and then slowly declines. Conduits that open initially as fissures are enlarged in those sections where the flux is greatest and become inactive where the flow is least. In this way, the conduits take on more pipe-like forms. Later, as flow rates decline, vents become increasingly constricted by solidification on their walls. A well-documented example is the Mexican volcano Paricutin. The eruption began with a small fissure and grew in strength for a few weeks as the vent became more centralized. During the following nine years from 1943 to 1952, the rate of discharge declined exponentially as the deep reservoir was slowly evacuated. Similar changes were observed during the eruption of the Icelandic volcano Hekla in 1947, 1948, and 1970 (Fig. 5.3). The eruption of 1970 broke out along a 25-km-long fissure but quickly became concentrated in a few central vents. As the pressure driving the magma declined, the walls of the conduit gradually closed until the flow was finally arrested. The rate of discharge at the beginning of the eruption was about 7,500 m^3 per second but declined rapidly to only a few cubic meters per second. The rates of discharge from large, central-vent volcanoes tend to be less than those from fissures.

5.5 ▲ Flood Basalts

On a number of occasions in the geological past, thick, widespread accumulations of basalt have been produced by giant fissure eruptions (Table 5.1). No other form of basaltic volcanism even approaches the

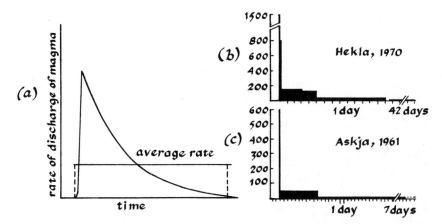

Fig. 5.3 (a) Variations of the rate of discharge of magma during a typical basaltic eruption. The average rate, which is equal to the total volume emitted divided by the duration of the eruption, is indicated by the horizontal line. Examples include **(b)** the 1970 eruption of the Icelandic volcano Hekla and **(c)** the 1961 eruption of Askja, also in Iceland. The rates for Hekla and Askja were reported by Thorarinsson and Sigvaldason in 1972 and 1962 respectively. (**a,** After Wadge, G. 1981. *J Volc Geoth Res* 11:139–68.)

TABLE 5.1

Ages and Aerial Extents of Some of the More Important Flood Basalts

Province	Age	Original Dimensions, km²
Lake Superior	Precambrian	125,000
Siberia	Permo-Triassic	2,500,000
Karoo, S. Africa	Jurassic	2,000,000
Parana, Brazil	Cretaceous	2,000,000
Ontang-Java	Cretaceous	1,500,000
Brito-Arctic	Eocene	200,000
Deccan, India	Eocene	500,000
Columbia River	Mid-Miocene	163,000

Note: The original dimensions of pre-Tertiary examples are very conjectural. The submarine Ontang-Java Plateau is in the Southwestern Pacific between New Guinea and the Marshall Islands. The Brito-Arctic Province encompasses parts of Eastern Greenland, Ireland, Scotland, and the Faeroe Islands.

scale of these outpourings. The most familiar example in America is the group of basalts forming the Columbia River Plateau of Oregon and Washington (Fig. 5.4). Hundreds of lava flows, some with volumes exceeding 2,000 km³, covered about 163,000 km² and had a total volume of about 174,300 km³. Yet the Columbia River basalts rank

Fig. 5.4 The flood basalts of the Columbia River Plateau were erupted from fissures in eastern Oregon and Washington and flowed as far as the Pacific Ocean. (Photograph by A. R. McBirney.)

only 10th in size among flood basalts of the world. The Mesozoic Karoo basalts are roughly 12 times larger, both in areal extent and volume. Although oceanic ridges produce greater volumes of basalt, this is only because their activity is more prolonged; eruptions of flood basalt are restricted to shorter intervals of time. By far, the greatest volume of basalt in the Columbia River group was erupted within a period of only 2 or 3 million years during the mid-Miocene. Individual eruptions were separated by periods of repose lasting thousands of years, but once initiated, their rates of discharge were of the order of tens or hundreds of cubic kilometers per day.

Flood basalts are not confined to the continents. Oceanic flood basalts of Cretaceous age cover a large region of the southwestern Pacific Ocean between New Guinea and the Marshall Islands. The largest group forms the Ontong-Java Plateau, covering an area of 1.5 million km² with a relief of about 3 km above the normal depth of the seafloor. Their vent systems cannot be seen, but it is possible that they were discharged from intraplate fissures.

Large fissure eruptions of basalt are found in a variety of tectonic settings. Some were associated with continental rifting and opening of oceanic basins. The basalts along the eastern coast of Greenland, for example, correspond to similar flood lavas in northern Ireland and Scotland and date from the opening of the North Atlantic in Eocene time. Others, such as those of the Lake Superior region and Siberia, seem to have been associated with rifting that was less extensive or aborted at an early stage. The basalts of the Deccan Plateau of India are thought to be the initial expression of a hotspot over which the lithosphere has moved northward, so the center of activity is now under the active volcano Piton de la Fournaise on the island of Reunion in the western Indian Ocean.

The closest historic analogs to these flood basalts are the fissure eruptions of the Icelandic volcanoes Eldgja and Laki. The first occurred around AD 940 and produced about 17 km³ of basalt; Laki discharged a similar amount in 1783. The Laki eruption is the better documented of the two. Following a week of earthquakes, magma began to issue from the fissure on the 8th of June and did not end until eight months later. The type of activity varied along the 25-km length of the fissure. Explosive eruptions predominated near the ends, where little or no magma was discharged, but explosions hurled out abundant lithic debris, and at least two cones of well-stratified ash were built by explosive eruptions where rising magma entered water-soaked ground. Nearer the middle of the fissure, where explosive activity was most intense, many cinder cones developed; most are about 40 m high, but a few rose to 90 m. The maximum outflow of lava was also concentrated near the middle of the fissure, but the vents were generally confined to short stretches where the flows were associated with abundant spatter cones built by lava fountains.

All told, the Laki lavas cover 565 km² to distances as far as 25 km. Their volume was 12.3 km³, whereas the volume of pyroclastic ejecta was only 0.85 km³. Although the flows were exceptionally fluid, almost all are blocky and scoriaceous. During the first 50 days, the average rate of emission exceeded 5,000 m³ per second. The huge amounts of carbon dioxide and aerosols put into the atmosphere and their effect on the global climate are discussed in Chapter 13.

5.6 ▲ Basaltic Lavas

As they move, very fluid lavas develop a smooth, ropey crust, often referred to by the Hawaiian name *pahoehoe* (Fig. 5.5). In many instances, lavas of this kind form a solid roof over their fluid interior, so the lava flows within a thermally insulated channel. Because the heat loss is greatly reduced, lavas fed by these tubes travel much farther than those in open channels. When the lava is drained, it leaves a *lava*

Fig. 5.5 Examples of **(a)** pahoehoe, **(b)** aa. The examples of pahoehoe and aa are from the Galapagos Islands.

continued

(c)

Fig. 5.5 continued Examples of **(c)** blocky lava. The example of blocky lava is from the Cascades. (Photographs by A. R. McBirney.)

tube some tens or hundreds of meters long. Fluid flows of this kind release a large part of their volatile components at or close to the vent. The result is a spectacular display of fountaining (Fig. 5.6) that builds *spatter cones* and *ramparts* consisting of agglutinated bombs and small droplets or filaments of glass called *Pelé's tears* and *Pelé's hair,* after the Hawaiian goddess of fire.

With increasing viscosity, basaltic lavas spread more slowly and come to rest at shorter distances and on steeper slopes. Their scoria-ceous surface forms what is often referred to as *aa* (Fig. 5.5b), another term of Hawaiian origin. They consist of jumbles of scoriaceous frag-ments of all sizes mantling a more massive interior. Because the mar-gins of a flow cool and solidify more quickly than the interior, they form levees of rubble between which the moving lava continues to ad-vance. Being more viscous, aa lavas require steeper slopes to move away from their source, and if the slope is gentle, they form steep-sided mounds over their vent.

Both pahoehoe and aa lavas may be present in a single flow, but al-though a pahoehoe flow can change to aa, the reverse is rarely seen. The change usually comes as the lava cools and loses volatiles. At its source, however, the lava may change from aa to pahoehoe or vice versa. For example, during the eruptions of Mauna Loa in 1859 and

(a)

(b)

Fig. 5.6 (a) Fountaining during the opening phase of a fissure eruption on Kilauea. The caldera can be seen behind the fume cloud. **(b)** A lava tube on the flanks of the Kilauea volcano in Hawaii. Note the strand lines on the wall marking the level at which the lava once flowed. (Photographs courtesy of the Hawaiian National Park Service.)

again in 1880–1881, aa lavas were discharged following an initial stage of strong fountaining in which large amounts of gas were released. As the fountaining subsided and the rate of discharge of lava diminished, the new flows retained more volatiles and took on the form of pahoehoe. The rate of flow is also a factor. The change from pahoehoe to aa is accelerated when the magma undergoes strong internal shearing. Thus, although they are more viscous, aa flows may be discharged at greater rates than those of pahoehoe, and for a given slope, the rate of advance of aa flows may be greater. This apparent anomaly is due to the effect of the rate of shear on the transition from pahoehoe to aa. Aa lavas do not flow faster because they are aa; they are aa because they flow faster.

Small "rootless craters" are a common feature of basaltic lavas. Some of these cones *(hornitos)* are a result of an explosive release of gas trapped in the lava; others are formed when lava passes over water-saturated ground. Many *littoral cones* near the shores of oceanic islands are deceptively similar to normal vents, but they are caused by steam explosions disrupting the lava when it flows over water. If fed by a well-established channel or lava tube, these cones may continue to erupt until the lava ceases to flow.

5.6 ▲ Vesiculation and Jointing of Lavas

An important process involved in the emplacement of lava flows on low-angle slopes is that of flow inflation. During inflation, lava sheets increase in height by an internal addition of lava. The crust of inflated flows grows as the cooling front propagates downward from the upper surface. Bubbles are trapped by the growing crust to form *vesicles* that define the extent of crust that grew during emplacement of the lava. After flow ceases, bubbles remaining in the melt rise and coalesce to form large "bell-jar" vesicles at the top of the still-fluid core. Thus, the internal structure of an inflated flow is distinguished by an upper vesicular zone (the upper crust) formed during emplacement, a dense core, and a thin vesicular zone at the base formed when cooling was sufficiently rapid to trap bubbles before they rose into the interior. The thickness of the upper vesicular zone can be used to estimate an emplacement time based on the measured rate of conductive cooling of inflated flows in Hawaii:

$$C = 0.0779 \ t^{1/2}$$

where C is the thickness of the crust in meters, and t is the length of time in hours.

After coming to rest, cooling lavas develop fractures and joints, many of which take on the form of hexagonal columns (Fig. 5.7). Oriented perpendicular to the cooling surface, columns tend to be vertical in horizontal lavas and horizontal in vertical dikes. Passing from crudely developed joints near the boundaries, they become more regular toward the interior where cooling was slower. In this way they often develop two sets of columns, one advancing upward from the base and another downward from the top. Joints are most regular in slowly cooled basaltic lava (Fig. 5.8), but they have been observed in lavas of virtually all compositions. The eroded surface of a flow with regular columnar joints resembles a surface paved or tiled with blocks a

Fig. 5.7 In very thick lavas, the complex jointing can be divided into several distinct zones. Just above a base of vesicular breccia is the *lower colonade*, which is overlain by thinner, irregular columns of the *entablature*. The *upper colonade* has larger columns that become more vesicular upward. (Aldeyarfoss, Iceland; Photograph by J-M Bardintzeff).

Fig. 5.8 (a) Columnar basalts at Devil's Post Pile, California. **(b)** A pavement of columnar basalt at the Giant's Causeway, Ireland. (**a,** Photograph courtesy of the National Park Service; **b,** photograph by Rohleder.)

few tens of centimeters across. The famous Giant's Causeway in County Antrim, Northern Ireland (Fig. 5.8b), owes its name to a remarkably regular pattern of this kind.

5.7 ▲ Lava Lakes and Ponds

A basic distinction can be made between *lava lakes* and *lava ponds*. The former are continually fed at their base by fresh magma, whereas the latter result from lava flowing into a topographic depression. Lava lakes remain molten for long periods, whereas ponded lavas immediately begin to cool and solidify.

Persistent lava lakes (Fig. 5.9) are maintained by fresh, hot magma that cools, degasses, and descends again to a deeper reservoir somewhere beneath the volcano. The incoming magma forms small fountains where it enters the lake and releases gas. A crust then forms on the surface as convection slowly carries the lava toward the margins of the lake where it descends and returns to its source. In this way the heat content of the lake is sustained, and the lake level can remain

Fig. 5.9 The lava lake of Kilauea in 1920. Magma rises from one or more sources in the interior of the lake and then moves outward toward the rim. Losing heat and gas, it forms a thin crust before descending and returning to a reservoir somewhere beneath the lake. In this way the level of the lake remains more or less constant for weeks or months. From time to time, however, the lava may overtop the rim, or it may drain back to its source. (After Jaggar, T. A. 1947. *Soc Am Mem* 21; 508 p.)

nearly constant for long periods. The amount of heat and gas given off by the lake is large relative to the amount of lava exposed at any one time. It is difficult to explain how lava lakes can remain active so long, but they must be sustained by a large subsurface body of magma, possibly in dikes.

The "fire pit" of Halemaumau was a permanent feature of Kilauea volcano from 1823 until it suddenly disappeared in 1924; it reappeared for 33 days in 1934, for 136 days in 1952, and again for 251 days in 1967. The lava lake in Nyamuragira, Zaire, began around the beginning of this century and disappeared in 1938; that of Nyiragongo, also in Zaire, was active from 1928 until 1977 and returned in 1982. Other lava lakes, such as those of Erta Ale (Ethiopia), Erebus (Antarctica), and Masaya (Nicaragua) have been active for decades or even centuries.

The lava lake of Nyiragongo is remarkable for the large changes of level that occur from time to time. Between 1959 and 1976 it rose 200 m, but on September 10, 1977, following a violent eruption, it was drained by eruptions through lateral fissures. Under the pressure of the heavy column of magma, lava broke out on the flanks of the volcano at velocities of nearly 100 km/hr. Then on June 21, 1982, the lake returned with an influx of as much as 2.5 million m^3 per day and by October of that year had formed a body of magma 500 m across and 400 m deep (Fig. 5.10).

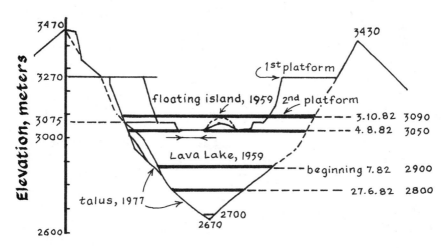

Fig. 5.10 Schematic cross section through the lava lake of Nyiragongo, Zaire, showing the successive levels reached at different dates during 1982 compared with that of 1959. The two benches around the lava lake of 1959 collapsed into the vent in 1977. (Modified from Tazieff, H. 1984. *J Volc Geoth Res* 20:267–80.)

Halemaumau, the lava lake that from time to time has been a persistent feature of the Hawaiian volcano Kilauea, has also dropped quickly by hundreds of meters, usually when an eruption has broken out on the lower flanks. Lava lakes were observed at the time of the eruption that began on January 3, 1983, and still continued 16 years later after having discharged more than 1.5 km³ of lava. One of these lava lakes, Kupaianaha, appeared in 1986 and disappeared during the eruptive phase of April and May of 1990. Another picked up the activity at the beginning of 1991 in the crater of a 250-meter-high adventive cone called Pu'u O'o.

While they are active, lava lakes provide an unexcelled opportunity to collect gases and measure the physical properties of fresh magma. Some of the best analyses of magmatic gases come from samples collected from the lava lakes of Halemaumau and Nyiragongo (Table 3.1).

Lava ponds take a variety of forms depending chiefly on the shape of the depression into which they flow. Thick accumulations of lava that fill craters or other deep depressions remain fluid for years or even decades. Three ponded lavas have recently been formed on the Hawaiian volcano Kilauea: Kilauea Iki in 1959 (120 m deep), Alae in 1963 (15 m deep), and Makaopuhi in 1965 (83 m deep). As they slowly cooled and solidified, these bodies provided exceptional opportunities to study the slow crystallization of magma. By drilling through the crust and into the partly molten interior, it has been possible to record the temperature distribution, growth of crystals, and changing composition of the remaining liquid (Fig. 2.4).

The time required for ponded lavas to solidify completely is mainly a function of their ratio of depth to surface area; it has ranged from 10.5 months for Alea to 25 years for Kilauea Iki. Solidification advances simultaneously from the top down and bottom up but is more rapid from the surface where heat is lost by radiation and to rainfall; at the base, conduction of heat to the underlying rocks is much slower. The solid crust on the surface of Kilauea Iki had a thickness of 20 cm after the first day, 2 m after a month, and 8 m after a year. In 1976, 17 years after its formation, the lava had solidified to a depth of 45 m. The rate of decline of temperature is relatively slow as long as liquid continues to crystallize but accelerates once the body is totally crystalline (Fig. 5.11). At the same time, viscosity increases at an accelerating rate. Between a temperature of more than 1200 and 1070°C, the measuring instruments sank freely into fluid lava. Between 1070 and 980°C, the partly molten lava was more rigid, and below 980°C it was completely solid.

In a few instances, lava has been ponded on relatively flat ground. During the eruption of Mauna Ulu, Hawaii, in 1974, lava spread slowly over a relatively flat area, and cooling of the margins produced a wall

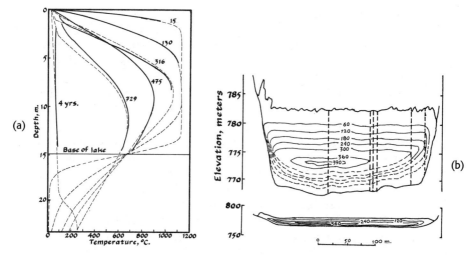

Fig. 5.11 (a) Evolution of temperatures in the lava lake of Alae, Hawaii, as a function of depth and time. The temperatures shown by solid lines were measured in boreholes drilled at intervals of time from the number of days indicated on each curve. The dashed lines are temperature profiles calculated for conductive heat loss; they have an average difference of less than 2°C from the measured values. **(b)** Longitudinal section through the lava lake of Alae showing the changing position of the 1000°C isotherm as the lens of melt became smaller with time. The locations of boreholes are indicated by dashed vertical lines. The numbers on the contour lines give the number of days elapsed since the lake was formed on August 22, 1963. The contours are dashed where they are based on extrapolations. In the upper figure, the vertical scale is 10 times greater than the horizontal. In the lower figure, the horizontal and vertical scales are equal. (Modified from Peck, D. 1978. *US Geol Surv Prof Paper* 935B.)

enclosing a more or less circular lake of molten lava. As more lava entered the area, it occasionally overflowed and heighten the enclosing banks. Heat losses from the broad surface area of "perched" lakes of this kind limit their size to a few hundred meters in diameter.

5.8 ▲ Viscous Lavas

Almost all andesites and other viscous lavas are of the aa or blocky variety (Fig. 5.5c). In contrast to the low shields produced by fluid basaltic lavas, most volcanoes built by more siliceous lavas have high central peaks, much steeper slopes, and greater proportions of fragmental material. The andesitic cones and domes of the High Cascade Range are familiar examples (see Chapter 10).

One of the best studied eruptions of viscous lava is that of the Costa Rican volcano Arenal. Starting in 1968, its summit vent discharged essentially continuous flows of aa and blocky lava at moderate temperatures ranging from 1100 to 1150°C. Viscosities calculated from measurements made 50 to 200 m from the crater are of the order of 10^7 Pa S (or 10^8 poises). This relatively high value results from their advanced crystallinity and moderately large amounts of silica.

The eruptions of lava evolved in three successive stages (Fig. 5.12). (1) During the initial stage, the emerging magma produced a flow 25 to 30 m wide and 10 to 15 m thick; it advanced at a rate of 50 to 70 m per day, depending on the nature of the terrain over which it moved. (2) As the mature flows moved down the flanks of the cone, they were 50 to 80 m wide, but on reaching the base they fanned out to widths of 100 to 200 m and their thicknesses increased to 15 to 30 m. A few were 400 to 500 m wide and 100 m thick. (3) In the final phases, the discharge of lava declined and eventually stopped, but as long as the interior remained relatively fluid, the flow continued to advance, and the crust slowly subsided. A "wave of subsidence" progressed down

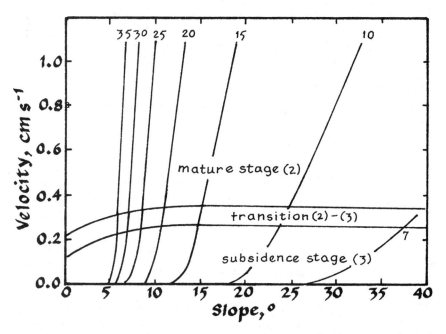

Fig. 5.12 The relations of the velocity of lavas of the Arenal volcano in a channel to the angle of slope for different thicknesses of the flow (numbers on curves in meters). Note that the velocities decrease from the mature stage (2) to the stage of subsidence (3). (1 cm s⁻¹ = 36 m hr⁻¹.) See text for details. (Modified from Cigolini, C., A. Borgia, and L. Casertano. 1984. *J Volc Geoth Res* 20:155–76.)

the slope at a rate of about 50 m per day. Fig. 5.12 shows how the velocity of the flow varied with the slope angle during the second and third stages.

The most viscous flows from Arenal consisted entirely of angular blocks, each having a solidified surface and a semimolten interior. As the flows advanced, these blocks rolled down the steep leading edge and were overrun as the flow advanced. The flows also carried blocks of older rock torn from the flanks of the volcano somewhat in the way glaciers pick up the rocky debris they leave in moraines. Some of this material became incorporated into the interior of the flow or, if very abundant, formed masses of breccia, one at the base and another at the surface.

During the 16-year eruptive episode between 1968 and 1984, Arenal discharged 51 flows with an average length of 2 km. In the first five years of this period the lavas were intermediate between the aa and blocky types. Later, they became blocky and took on a bimodal character, with various proportions of blocks and liquid. About 15% of the blocks had sizes of about 1 meter; 80% were between 30 and 60 cm, and the remaining 5% ranged in size down to a few millimeters. The entire mass moved slowly down the slope as the force of gravity worked against the internal friction and resistance of the rough ground surface. Where the resistance to flow was greater than a certain threshold value, the flows tended to split and become discontinuous, while the flow fronts became higher and steeper.

Although very siliceous lavas are not uncommon in the geological record, no eruption of rhyolitic lava has been observed by a geologist. Judging from their morphology, most rhyolitic flows are very viscous and move only short distances from their vents, but a few, such as those of southern Texas, are large and remarkably extensive. They must have been erupted at temperatures near or even above their liquidus.

The surface morphology of viscous lava is a function not only of its rheological properties (viscosity and yield strength) but also of the rates of discharge and cooling. The slower the rate of effusion the rougher the surface. Slowly discharged lavas and those that are rapidly cooled have very irregular surfaces with steep-sided spines that project through a carapace of large, jumbled blocks.

5.9 ▲ Interpreting Older Lavas

As weathering and erosion strips away the surficial features of lavas, it becomes increasingly difficult to interpret their form and eruptive history. When flows are piled one atop another in thick sections, one sees them only in discontinuous exposures, such as stream valleys and shorelines. Sorting out individual flows and identifying their ages, sources, and extents are challenging tasks that require all the tools at a geologist's disposal.

One can usually define the tops and bottoms of individual flows by means of their scoriaceous or rubbly surfaces, but it is important to bear in mind the possibility that overlapping tongues from a single flow may be separated by surfaces that represent time intervals of only hours or days. A more reliable marker is the layer of soil that develops on the weathered surface of older flows. These soils are commonly baked to a brick-red color and may contain organic material that one can date by the carbon-14 method.

Because massive lavas are relatively resistant to erosion, they commonly cap flat-topped mesas or table mountains. Fossil lava ponds are

Fig. 5.13 Devil's Tower in Wyoming is a spectacular example of differential erosion of a volcano that had a large ponded lava in its crater. The weaker rocks on the flanks have long since been removed by erosion. (Courtesy of Devil's Tower National Monument.)

rather common. Devil's Tower in Wyoming and the Hopi Buttes of Arizona are spectacular examples. The lavas that filled craters, being more resistant to erosion than the weak debris on the flanks, form steep-sided hills above the surrounding topography (Fig. 5.13). Where the lava has flowed down a valley between weaker sedimentary rocks, subsequent erosion can produce an inversion of relief in which a lava flow that originally occupied low levels is perched at a higher elevation than later flows that came down newly carved valleys (Fig. 5.14).

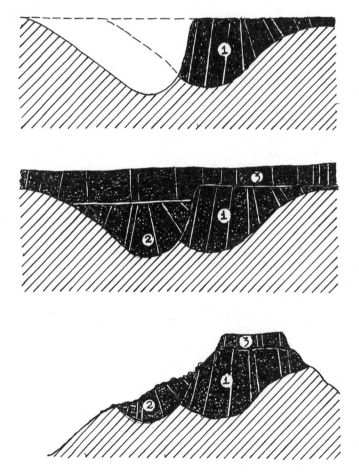

Fig. 5.14 Differential erosion can cause an inversion of topographic relief in which valleys become ridges capped by resistant lavas. When successive lava flows have flowed down a persistent drainage channel, erosion following each flow tends to follow the margins of the resistant basalt. Later flows, such as 2, may lie at a lower elevation than earlier ones and produce an apparent stratigraphic order (2-1-3) that differs from the true order of eruption, 1-2-3.

Unless exposures are almost continuous, individual flows can rarely be traced to their sources, but distinctive geochemical features may make it possible to "fingerprint" a particular lava and match it with a possible eruptive vent. For example, the lavas of the Columbia Plateau have been grouped into a number of sets on the basis of subtle compositional differences, and this has made it possible to relate them to different groups of dikes that served as their feeders.

Lavas with similar petrographic and chemical features may have the appearance of single flows of great areal extent when, in fact, they have come from different sources. Repeated eruptions from numerous scattered vents over relatively short periods can lead to extensive "plateau lavas" covering vast regions. The lavas covering the High Plains of Central Oregon are a good example. They differ from flood basalts in that the latter are much more voluminous and flow for greater distances from a few fissure sources in a restricted locality.

Suggested Readings

Hulme, G., 1974. The interpretation of lava flow morphology. *Geophys J Roy Astr Soc* 39:361–83.
A classic study of the physical behavior of moving lava.

Kilburn, C. R. J., and G. Luongo, eds. 1993. *Active lavas.* London: UCL Press, 374 p.
An exceptionally comprehensive treatment of the broad spectrum of lava flows from many localities around the world.

Reidel, S. P., and P. R. Hooper, eds. 1989. Volcanism and tectonism in the Columbia River Basalt Province. *Geol Soc Am Spec Pub No 239*, 386 p.
An excellent compilation of information of the flood basalts of eastern Washington and Oregon.

Swanson, D. A., T. H. Wright, and R. T. Helz. 1975. Linear vent systems and estimated rates of magma production and eruption for the Yakima basalt of the Columbia Plateau. *Am J Sci* 275:877–905.
A thorough study of the Miocene eruptions of flood basalts in eastern Washington and Oregon.

Tilling, R. I., and D. W. Peterson. 1993. Field observations of active lava in Hawaii: Some practical considerations. In C. R. J. Kilburn and G. Luongo, eds., *Active lavas*. London: University College London Press, pp. 147–74.
Quantitative observations on moving lava from the basaltic volcanoes of Hawaii.

CHAPTER 6

Domes

▼

6.1 ▲ Forms of Domes

When lava is so viscous that it can flow only a short distance, even on very steep slopes, it accumulates over its vent to form a steep-sided dome or spine. Growth may be either *endogenous,* when magma is injected and inflates the interior, or *exogenous,* when it accumulates as extrusions on the surface. Few domes grow by a single mechanism; most change from one form to another as they become larger. The dome of Mt. St. Helens (Fig. 6.1), for example, was mainly exogenous in its early stages of growth but became more endogenous with time. Almost all domes have some component of endogenous growth, and many are composite in the sense that they are made up of multiple, coalescing extrusions.

Many domes, such as the rhyolitic domes of the Coso and Mono Craters districts of California, grow in craters formed by an initial explosive eruption (Fig. 6.2). Others, such as Showa Shinzan (Fig. 6.3), begin as shallow intrusions that later break through to the surface.

Most domes are composed of silica-rich, viscous magmas, such as dacite, trachyte, or rhyolite, but even basic lavas may be extruded as domes if the temperature and gas contents are low enough. Domical lavas are characteristically rich in glass; indeed, many consist of obsidian. But in the cores of most large domes the glass is replaced by a microcrystalline groundmass. Although the blocky crusts may be pumiceous, coarsely vesicular lavas are exceptional. On the other hand, the lavas of many andesitic domes have a microporous texture formed by a mesh of interlocking crystals. The content of phenocrysts varies widely, from almost zero to more than 30%.

85

(a)

(b)

(c)

Fig. 6.1 (a and **b)** Steep-sided dacitic domes of Lassen Peak, California. **(c)** The dome formed in the crater of Mt. St. Helens following the eruption of 1980. (**a** and **b,** Drawing by H. Williams; **c,** photograph by A. R. McBirney.)

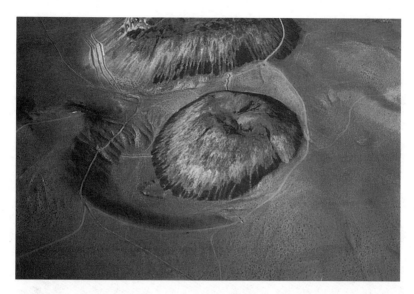

Fig. 6.2 A rhyolitic dome in the Coso volcanic field of southern California has grown inside a crater formed by an initial explosive eruption. (Courtesy of W. A. Duffield.)

6.2 ▲ Principal Types of Domes

A great number of terms have been coined to describe the various morphological forms of domes, but most fall into one of four or five main types (Fig. 6.4).

1. *Cryptodomes* are formed by shallow intrusions of magma that cause the ground surface to be upheaved, sometimes by hundreds of meters. A recent example is the dome of Usu-shinzan, Japan, that formed in 1977 and 1978 without ever breaching the surface. Intrusive domes of this kind are especially common in sedimentary basins. A spectacular example is the complex of Pliocene cryptodomes at Sutter Buttes, California. Rising bodies of viscous dacite up-arched the overlying sedimentary beds over an area 10 km across. Subradial faults were formed, breaking the sediments into sectors and tilting them outward at angles as steep as 60 degrees. In one sector, beds with an aggregate thickness of 1500 m were turned to verticality and locally overturned.

2. *Plug domes* are formed when very viscous magma is extruded from a vent like toothpaste from a tube. The domes of Lassen Peak illustrated in Fig. 6.1 are a well-known example. They produce steep-sided columns with a radius approximately equal

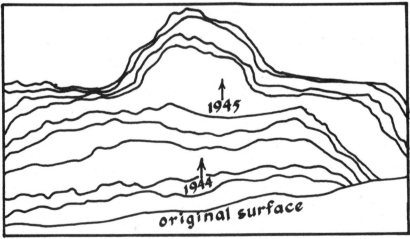

Fig. 6.3 The dome of Showa-shinzan began as a shallow intrusion that eventually broke through to the surface. The stages of growth are shown below.

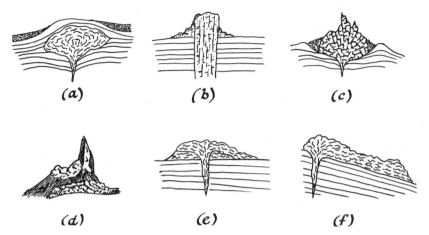

Fig. 6.4 The main types of domes. **(a)** Cryptodomes are produced by intrusions at a shallow depth below the ground surface. **(b)** Plug domes result from upheaval of very viscous lava at the surface, **(c)** Peléan domes are piles of blocky debris derived in large part from collapse of plug domes and from shallow intrusions. Spines **(d)** commonly protrude above their summit. **(e)** Lava domes grade into viscous lava flows. If extruded on sloping ground they take on an asymmetrical form known as a coulée **(f).**

to that of their vent. In rare cases they may form *spines* with heights of a hundred meters or more. The flanks of a plug dome or spine may be striated and polished by abrasion of the nearly solid magma against the walls of the conduit.

Many plug domes begin to grow as shallow intrusions and later breach the surface. The dome of Showa-shinzan in Japan, for example (Fig. 6.3), began to grow as a cryptodome in early 1944 and in the following months lifted the ground surface about 200 m before a central plug broke through to rise another 110 m.

3. *Peléan domes,* named for the domes formed on Mt. Pelée, Martinique, in 1902–1905 and 1929–1932, are a larger, more evolved form of plug domes that have crumbled into chaotic piles of talus. They tend to be irregular in form, especially if they form composite clusters, and may have a spine protruding through the blocky debris. Their coarse, blocky slopes give them a crudely conical shape with steep slope angles.

4. *Lava domes* are made up of overlapping extrusions of viscous flows extending only short distances from a central vent. The ratio of their height to width is less than that of the preceding types. They tend to be less symmetrical because their growth is more

strongly influenced by inclinations of the underlying ground surface and the different courses followed by successive flows.

6.3 ▲ Shapes and Sizes

The geometrical configuration of domes is best defined by the ratio of their height to width (Fig. 6.5a) and is mainly a function of the viscosity and rate of discharge of the magma. It can be approximated by the following simple mechanical model that assumes a symmetrical hemisphere surrounded by an apron of talus and resting on a flat base (Fig. 6.5b):

$$D = \sqrt{\frac{\sigma t}{\gamma}} / h$$

where σ is the strength of the semirigid crust, t is the thickness of the crust, γ is the magma's unit weight, and h is the internal pressure under the apex of the dome. The calculation treats the magma as a liquid under pressure. The value D given by this equation is a dimensionless number governing the ratio of height to width; in the case of Mt. St. Helens (Fig. 6.1b), it was approximately 1.0. Other variables, such as the viscosity of the magma and the physical properties of the carapace, have relatively less influence. As it grows, a dome is constantly changing its shape as slumping and explosions alter the regularity of its flanks, but between these events its form tends to approach the equilibrium profile.

Peléan spines, although very rare, are among the most spectacular of all volcanic features. Fig. 6.6 illustrates the gigantic spine that began to rise from the summit of Mt. Pelée in November of 1902. Between the 9th and 12th of November, it rose 60m, and by the 24th of November, it had reached a total height of 200 m and elevation of 1575 m before starting to crumble. It continued to rise through three successive stages, each ending with collapse into a pile of rubble. If it had remained intact, the spine would have been 850 m high and would have reached an elevation of 2000 m above sea level.

At Mt. St. Helens, the dome of 1980 had a volume of 4 million m³, but after nine phases of growth and explosive eruptions during 1981 and 1983, it attained a size of 44 million m³ and measured 880 by 830 m at its base and 224 m in height. Thereafter, it continued to grow but at a steadily decreasing rate. Similar rates were measured during the growth of Mt. Unzen (Fig. 6.7) between May of 1991 and May of 1995 and of Monserrat in the Antilles between September of 1995 and July of 1997. The rates of growth of closely observed domes have

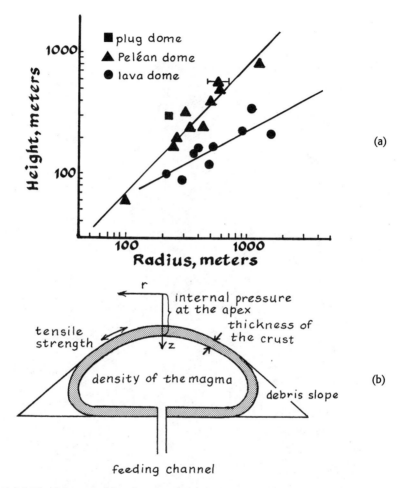

Fig. 6.5 (a) The relation of height to radius of different types of domes. Correlation lines do not necessarily project back to the origin; the line for Peléan domes *(triangles)* has a slope of 1.06, whereas the one for low-aspect-ratio lava domes *(solid circles)* is 0.55. Note that the single point on the diagram for an extruded plug dome (O'Usu, Japan) has a height-to-radius ratio greater than 1.0 *(solid square)*. **(b)** A theoretical model for the shape of a dome is based on the balance of four mechanical factors: the thickness of the crust, its tensile strength, the density of the magma, and its internal pressure. (From Blake, S. 1990. Viscoplastic models of lava domes. In J. Fink, ed., *Lava flows and domes*, Berlin: Springer-Verlag, pp. 88–126.)

Fig. 6.6 The spine of Mt. Pelée reached a height of 1,566 m on March 12, 1903. Note human figures in foreground. (From Lacroix, A. 1904. *La Montagne Pelée et ses éruptions*. Paris: Masson, 662 p.)

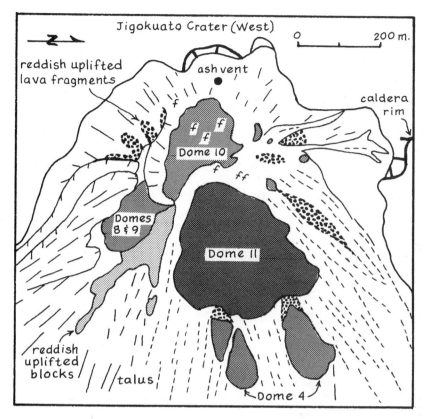

Fig. 6.7 Sketch map of the lava dome complex of Unzen, Japan, in early June 1993.
Fumaroles in dome 10 are indicated by the letter *f*. (From Nakada, S. 1993. *Bull Global Volcanism Network* 18[5]:8–9.)

ranged from 0.14 m³ per second at Galeras in Columbia in 1991 to
3.5 m³ per second at Unzen in 1991 to 5.2 m³ per second at Monsterrat
in June 1997.

Domes commonly grow in clusters that coalesce to form large complex structures. The eruption of Mt. Unzen in Japan, which began in
May of 1991, produced a succession of domes of this kind (Fig. 6.8).
In December of that year, dome no. 6 reached an elevation of 1,375 m,
then at the end of March and the beginning of April of 1993, domes 10
and 11 grew simultaneously. Despite frequent slope failures, dome 10
reached an elevation of 1,440 m, the highest point of the entire complex. It is estimated that between the 20th of May 1991 and the beginning of March of 1993, the output of lava, almost all of which contributed to the domes, was 130 million m³. By the time the eruption
ended in May 1995, 13 domes had been formed.

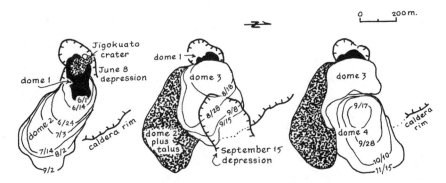

Fig. 6.8 Growth of successive domes on Unzen during the period of May through November 1991. Dome 1 formed on the margin of Jigokuato crater in early June. The numbers 6/1 through 9/2 indicate the edge of dome 2 at the dates June 1 through September 2. Dome 3 appeared on August 18 (8/18) and dome 4 on September 17 (9/17). The barbed line indicates the edge of the crater. (From Nakada, S., and T. Fujii. 1993. *J Volc Geoth Res* 54:319–33.)

6.4 ▲ Internal Structure

The internal complexities of domes are a reflection of the varied ways in which they grow. Fluidal banding is well developed in some domes and absent in others. It is usually concentric with the surface (Fig. 6.9), especially in intrusive domes, but it may also have a sheaf- or fan-like form.

Concentric banding is most characteristic of the central parts, directly above the feeding conduits, and usually changes to fan-shaped banding away from the center. Inward dips of the flow planes flatten at the margins and in endogenous domes, and lava mounds may even turn over and dip outward. Exfoliation of flow planes resembles the opening of petals on a flower bud expanding from within. Jointing is generally irregular or crudely radial and results largely from internal expansion, but some contraction joints are parallel to the surface and may easily be mistaken for flow planes.

A preferred orientation of elongated phenocrysts may indicate directions of movement, but this is a less reliable criterion of the direction of flow than striations on shear planes. Shearing along internal flow planes or against conduit walls can produce striated, polished surfaces, especially in plug domes; however, where the surrounding rocks are weak and friable, the margins consist of friction breccias with fragments ripped from the walls and incorporated into the sides of the dome.

Fig. 6.9 A section through a dissected lava dome on the island of Lipari, Italy, shows the well-developed concentric banding and radial joints. (Photograph by H. Pichler.)

6.5 ▲ Explosive Eruptions from Domes

Domes commonly appear following explosive pyroclastic eruptions, just as the present dome in the crater of Mt. St. Helens began to grow after the explosive phase of the 1980 eruption. Their subsequent growth is not always quiet, however. Although the viscous magma is emplaced slowly as massive intrusions and angular blocks of rubble, gas is an important component and, depending on how it is released, may affect the mode of emplacement. All domes give off gas through cracks, sometimes for many years after they have ceased to grow. This is especially true of endogenous domes. But when cracks formed by expansion of the brittle crust are filled by magma, their permeability is decreased and degassing is impeded. This greatly increases the possibility of explosive outbursts. Because the magma is so viscous, exsolved gases are confined under high pressures, and when trapped in the interior, they can be released by sudden failure of an oversteepened flow front. The volcano Merapi, Java, has had repeated eruptions of this kind (Fig. 6.10). The mechanisms of these dome-related eruptions will be examined in greater detail in Chapter 8.

Fig. 6.10 The dome of Merapi, Java, during its activity in August 1980. In the background, from left to right, are the volcanoes Sumbing and Sundoro and the Dieng Plateau. The walls of the crater in which the dome has grown can be seen in the right foreground and behind the dome. (Photograph by J. M. Bardintzeff.)

Suggested Reading

Fink, J. H., ed. 1990. *Lava flows and domes: Emplacement mechanisms and hazard implications*. Berlin: Springer-Verlag.
A collection of excellent papers giving a comprehensive summary of recent work on all types of volcanic extrusions.

Gorshkov, G. S. 1959. Gigantic eruption of the volcano Bezymianny. *Bull Volc* 20:77–112.
An excellent account of a powerful eruption from a dome complex in eastern Siberia.

Yanagi, T., H. Okada, and K. Ohta, eds. 1992. *Unzen volcano: The 1990–1992 eruption*. Fukoka: Nishinippon & Kyushu University Press, 137 p.
A well-illustrated and comprehensive account of the growth of domes and associated explosive eruptions at Unzen.

Pyroclastic-Fall Tephra

▼

7.1 ▲ Types of Pyroclastic Eruptions

All magmas contain some proportion of gas that exsolves as the magma approaches the surface. Under some conditions this simply produces vesiculated lava, while in others it disrupts the magma and discharges it explosively, either as pyroclastic flows (the subject of the next chapter) or as explosive, high-angle ejections. Despite their great diversity (Fig. 7.1), explosive eruptions can be divided into two general types. During most basaltic eruptions, solid or plastic fragments of magma are ejected from the vent at high angles and, following a parabolic course through the atmosphere, blanket the surrounding topography with debris of all sizes. In other eruptions, smaller particles, ranging in size from millimeters to microns, are carried to heights of hundreds or thousands of meters where winds disperse them over wide regions. Eruptions of the first type in which incandescent fragments of all sizes follow ballistic trajectories are referred to as *Hawaiian* or *Strombolian*, depending on the strength of the explosion. The second type in which small fragments are propelled in a vertical, gas-rich eruption column is characteristic of *Plinian* activity. Weaker Plinian, or *sub-Plinian*, eruptions grade into a *vulcanian* type in which the eruption column does not reach great heights and the discharge of ash-laden gas is less continuous. Vulcanian eruptions grade into the *Surtseyan* type as more surface water contributes to the explosions. All these types of eruptions produce *tephra*, the general name given to all types of pyroclastic material. Unlike pyroclastic flows, which tend to follow topography, ash-fall tephra is deposited uniformly over hills and valleys alike (Fig. 7.2).

(a)

(b)

Fig. 7.1 Examples of the main types of explosive eruptions. **(a)** Typical fire fountaining during a rift-zone eruption in Hawaii. **(b)** A Strombolian eruption from
a newly formed volcano in eastern Siberia. In this case, a lava flow is being
discharged from the base of the cone, but this is not an essential part of this type
of eruption. The name is taken from that of the Italian volcano that erupts in
this way. **(c)** Vulcanian eruptions, such as this example from the volcano Barcena
off the coast of Mexico, are characterized by relatively low temperatures and
clouds of fine debris. **(d)** Surtseyan eruptions get their name from a volcano that
rose from the sea off the southern coast of Iceland. (**a,** Courtesy of Hawaiian National Park Service; **b,** photograph by G. S. Gorshkov; **c,** photograph courtesy of
U.S. Air Force; **d,** photograph by S. Einarsson.)   *continued*

(c)

(d)

Fig. 7.1 continued For legend see previous page.

(e)

Fig. 7.1 continued (e) Plinean eruptions, such as that of Mt. Mazama portrayed in this painting by Paul Rockwood, form columns of gas and ash ascending to heights of many kilometers. They are often the prelude to collapse of a caldera as was the case when this eruption resulted in what is now Crater Lake, Oregon.

Fig. 7.2 For legend, see next page.

7.2 ▲ Fragmentation of Magma

Several factors govern which of these forms an explosive eruption will take. As bubbles nucleate and expand, they begin to contact one another and coalesce, and, if abundant enough, they disrupt the continuity of the magma. Although the basic principles governing the disruption of magma are rather simple, the process is difficult to quantify, mainly because of the wide variety of factors governing eruptive behavior.

The total proportion of gas is obviously important, but so too are the viscosity of the liquid and the rate at which it rises. When gas can expand freely in a rising magma of low viscosity, it is largely decompressed when it emerges from the vent. This does not prevent it from being discharged at a high velocity, however. As vesiculation and expansion begin high in the column, the inflated gas–liquid mixture rises at an accelerating rate, much in the manner of a geyser. The spectacular fountaining during basaltic eruptions on Hawaii is a good example (Fig. 7.1a). On the other hand, if the proportion of gas is relatively small, it begins to separate and expand at a higher level and the velocity of ejection is not as great.

The level at which gas exsolves in a rising column of magma depends on two factors: the amount of gas in solution and on how easily bubbles nucleate and coalesce. The greater the proportion of dissolved volatiles, the deeper the level at which they will reach their solubility limit. A certain amount of oversaturation is required for bubbles of gas to nucleate, but it is normally less that about 10 bars, especially if CO_2 is an important component or if the magma contains crystals on which the bubbles can nucleate. Depending on the amount and composition of their volatiles, most magmas begin to vesiculate at depths of 1 or 2 km, but because CO_2 is much less soluble than water, it exsolves at greater depths. Basalts erupted on the seafloor, for example, are under such pressure that they are able to exsolve little water vapor, but they form bubbles of CO_2 quite readily. The solubilities of volatiles are not independent of one another; exsolution of CO_2 facilitates exsolution of other gases as well because the presence of

---

Fig. 7.2 As seen in this example from Oshima, Japan, ash-fall tephra is deposited with nearly uniform thickness regardless of the topography. This "mantle bedding" is in contrast to deposits of pyroclastic flows, which tend to follow valleys. When seen in older rocks, draped beds such as these have been misinterpretted as folding. Note that rapid erosion can produce marked unconformities, even during the course of a single eruption.

bubbles reduces the level of oversaturation required for them to come out of solution.

In contrast to basalts, bubbles in more viscous magmas expand more slowly and tend to be smaller when they approach the surface (Fig. 7.3). Because the gas is still compressed, it is released more violently when the magma is disrupted high in the vent. Thus, it imparts a rapid, cannon-like acceleration of the gas–magma mixture. This is in contrast to gas in less viscous magmas, which because it expands more freely deep in the column, is under less pressure on reaching the surface. (The role of external meteoric water in Surtseyan eruptions is quite different; it will be discussed in Chapter 10.)

A silicate melt may become greatly inflated without being totally disrupted. In many magmas, especially very siliceous ones, bubbles can grow and touch one another without coalescing; they are separated by a thin wall, so that the gas–liquid mixture becomes a foam. The cause of this difference lies mainly in the surface tension between the bubbles and liquid. Just as in soap bubbles or the foam on beer,

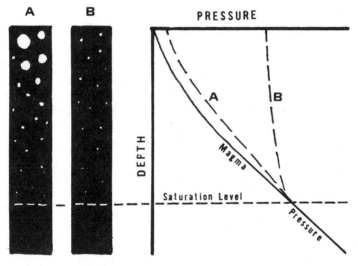

Fig. 7.3 The eruptive behavior of magma is strongly dependent on the manner in which gas expands as the magma rises. Assume two columns, A and B, are identical in all respects and begin to vesiculate at the same depth but, because of differences of viscosity or rate of ascent, the expansion in A is more rapid. As shown in the diagram on the right, pressure released at the surface will be greater for column B, and for that reason, that column will erupt with greater violence. The greater expansion in A would cause more acceleration, but the resulting fountaining would not be as explosive.

the surface tension adds strength to the thin walls between adjacent bubbles and impedes their coalescence. Another factor is the rate of expansion. Silicate liquids have characteristic relaxation rates at which their structures can respond to deformation. When deformed at rates more rapid than that at which they can flow, they rupture like glass. Thus, rapidly expanding bubbles are more likely to coalesce than those that grow more slowly. Relaxation rates are a function of several factors, the most important of which are probably silica and water content.

7.3 ▲ Sizes and Shapes of Pyroclastic Ejecta

Pyroclastic fragments are normally classified according to their size on a granulometric scale. The system shown in Table 7.1 distinguishes ash as being smaller than 2 mm, lapilli between 2 and 64 mm, and all fragments of larger size being called blocks or bombs, depending on whether their form is angular or rounded.

The morphology of ejecta (Fig. 7.4) is determined mainly by the viscosity and rate of cooling of the magma. Hot, fluid magmas tend to form rounded bombs, whereas more viscous or rapidly cooled ejecta are more angular. Bombs of fluid magma that cool slowly remain plastic and are affected by aerodynamic forces during flight. If still soft and plastic when they strike the ground, they flatten into what are called spatter or "cow-dung bombs." Continued expansion of gas bubbles in the interior of a bomb causes the brittle surface layer to crack and take on the appearance of "bread crust." Other terms used to describe pyroclastic material include *scoria,* a type of ejecta resulting from strong vesiculation of gas-rich, mafic magma, and *pumice,* the product of extreme vesiculation of felsic magma. The basaltic equivalent of pumice, *reticulite,* can be equally porous.

TABLE 7.1	
Granulometric Classification of Volcanic Ejecta*	
Blocks and bombs	>64 mm (-6ϕ)
Lapilli	>2 mm (-1ϕ)
Ash	
Sand-sized	>1/16 mm (4ϕ)
Silt-sized	>1/256 mm (8ϕ)
Clay-sized	<1/256 mm (8ϕ)

Note: Blocks and bombs have similar sizes, but the former are angular and the latter rounded.

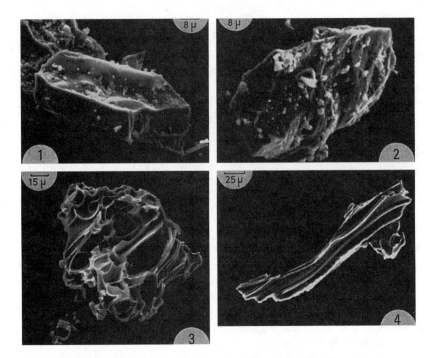

Fig. 7.4 Morphology of glassy tephra as shown by scanning electron microscopy. *1,* Angular shard of Santiaguito, Guatemala. *2,* Angular shard from Merapi, Java. *3, 4,* and *5,* Pumiceous glass from Mt. St. Helens, 1980. The siliceous water-saturated melt formed pumice with many fine, tubular cavities when the magma was very viscous and at temperatures below about 850°C, but at higher temperatures the magma was less viscous and formed shards as in *5. 6,* Inflated glass from the 1956 eruption of Bezymianny, Kamchatka. This rounded form with smooth surfaces results from the strong surface tension of the low-viscosity melt. Exsolution while the magma is still relatively fluid produces disrupted bubbles resembling miniature volcanoes. (Sample collected by G. Bogoyavlenskaya.) *7,* Pumiceous glass from the 1902 eruption of Mt. Pelée, Martinique. (Sample from the collection of A. Lacroix in the Museum of Natural History, Paris.) *8,* A spherical droplet of glass from the pumice blanket on Mont Dore, Central France, shows the effect of rapid cooling of magma that was just saturated with water and did not vesiculate on cooling. (Photographs by J. M. Bardintzeff and C. Jehanno.)

By combining the various terms for shape with those for size, one can speak of scoriaceous lapilli, pumiceous ash, bread-crust bombs, and so on. Even with all these descriptive terms, however, it is often difficult to provide a precise definition of material with a range of sizes. Distinctions based on the size distribution of particles provide

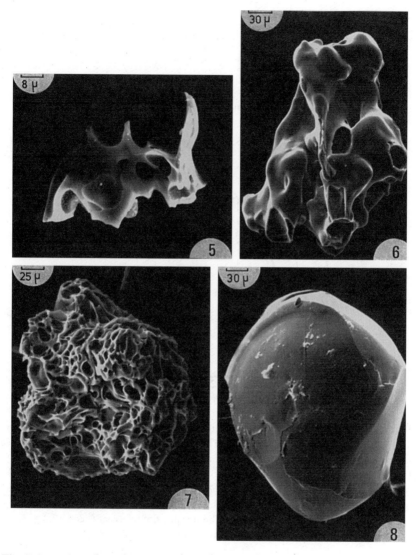

Fig. 7.4 continued For legend, see previous page.

more quantitative means of characterizing pyroclastic material. To determine the proportions of different sizes, samples of several kilograms are passed through a series of sieves to separate the particles according to a geometric progression in which each successive size decreases by half that of the preceding one. Each fraction greater than a given size is weighed, and its proportions are used to construct a cumulative curve of size versus weight fraction (Fig. 7.5a). These curves

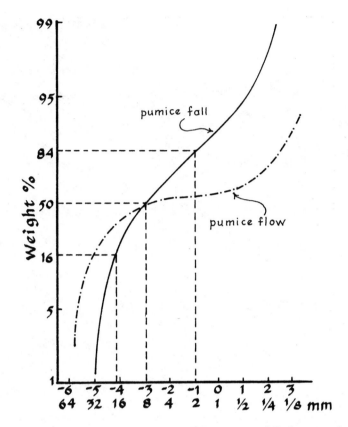

Fig. 7.5 Cumulative granulometric curves for pumice-fall deposits that were erupted by Mt. Pelée around AD 1300. The broken line is for an associated pumice flow (see Chapter 8). The median size at 50% is 8 mm (−3φ). The sizes at 84% (= −1φ) and 16% (= −4.2φ) are used to define the degree of sorting of the two deposits. The sorting coefficient, σφ = (φ84 − φ16)/2, is 1.6 for the pumice-fall deposits. The medium size is not very different for pumice-flow deposits (4.9 mm or −2.3φ), but the sorting coefficient for the pumice-flow deposit (σφ = (2.4 + 5.1)/2 = 3.75) indicates a much wider distribution of sizes.

normally have a sigmoidal form, but the shape for each type of tephra is distinctive.

Two useful parameters can be derived from this information: the median size and the degree of sorting. The median (Mdφ) represents the size at which the particles are divided into two equal parts, one larger and the other smaller than the median. The sorting coefficient (σφ) is the ratio (φ16 − φ16)/2, reflecting the homogeneity of the sample. Mdφ characterizes the general coarseness of the deposit, whereas σφ is a measure of the size distribution, with the value of σφ decreasing as the sizes become more uniform. Most ash-fall tephra is relatively well sorted. During transport, the heaviest and largest fragments drop out more quickly and reach the ground before finer particles. Thus, the deposits of a given ash fall tend to be stratified with the sizes and densities of particles decreasing both upward at a given location and laterally at a given stratigraphic level. The degree of sorting increases with the time the material is in the air and the distance it travels.

7.4 ▲ The Magmatic Budget of an Eruptive Column

The size and form of eruptive columns that rise during explosive events are a good measure of the intensity of Plinian eruptions. These columns can be divided into three parts, each distinguished by a different dynamic behavior (Fig. 7.6). The basal part of the column, usually a few hundred meters high, is characterized by high ejection velocities and rapid deceleration. The middle zone, which may be tens of kilometers high, is distinguished by intense turbulent convection. As the density of the expanding hot gas becomes less than that of the surrounding atmosphere, the column continues to rise by virtue of its thermal buoyancy. The upper limit of this zone is defined by the level at which the density of the column becomes equal to that of the atmosphere. The momentum of the column enables it to continue to rise above this level, but it soon loses velocity and begins to spread laterally, giving the column its typical mushroom shape. When the velocity at the level of the vent falls below about 100 m per second, the column

Fig. 7.6 Eruption columns can be divided into three zones according to height, velocity, and density, as shown by the curves on the right. The zone of convection corresponds to the interval in which the density of the eruptive column is less than that of air. (Adapted from Sparks, R. S. J. 1986. *Bull Volc* 48:3–15.)

collapses. Velocity is a function of the vent diameter and volume flux of the eruption, so magma must be discharged at a greater rate to sustain an eruption column from a large vent than from a smaller one.

The factors governing the form of an eruptive column (diameter, height, and velocity) depend on the magnitude of the event and the rate of discharge of magma, which in turn are functions of the duration and volume of the eruption. It is possible to determine these factors quantitatively from the form of deposits laid down by the eruptions, even when the event was not witnessed. The prehistoric P1 eruption of Mt. Pelée has been examined in this way.

The volume of pumice discharged during the eruption was about 0.95 km³, which, when divided by a factor of 5 to take into account the low density of this material (about 1.0) compared with that of lava (about 2.5) and the numerous voids that make up about 50% of the volume of ash, gives a value for an equivalent volume of 0.19 km³, dense rock equivalent (DRE).

The thermal energy liberated in the eruption has been calculated using the following formula:

$$E_{th} = V \cdot d \cdot T \cdot K \tag{1}$$

where E_{th} is in joules, V is the volume in km³ (in this case 0.95), d is specific gravity (1.0), T is the temperature of the ejecta (500°C), and K a constant including the specific heat of the magma and the mechanical equivalent of heat (8.37×10^{14}).

Thus, the thermal energy released would have been 4×10^{17} joules or about 20,000 times that of the Hiroshima atomic bomb. The kinetic energy would be much less, usually around a few percent of the thermal energy.

The erupted mass (0.19 km³) was made up of 85% (0.16 km³) pumice derived from new magma and 15% (0.03 km³) angular fragments coming from the older rocks evacuated to form a crater of this volume. The maximum value for the mass erupted is related to the rate of discharge according to the following equation:

$$M = \Pi \cdot r^4 \cdot d_c (d_c - d_m) \cdot g/8v \tag{2}$$

where r is the radius of the vent (60 to 70 m), d_c is the density of the crustal rocks (2.65×10^3 kg m⁻³), d_m is the density of the magma (2.5×10^3 kg m⁻³), g is the acceleration of gravity (9.81 m s⁻²), and v is the viscosity of the magma (5×10^5 Pa s).

The value obtained, 3.8 to 7.0×10^7 kg s⁻¹, corresponds to a volumetric rate of discharge of 1.5 to 2.8×10^4 m³ s⁻¹. Thus, the 0.16 km³ of new magma could have been erupted in about 1.5 to 3 hours.

The maximum velocity of ascent of the magma would be:

$$u = M/d_m \, \Pi r^2 \tag{3}$$

which yields values of 1.3 to 1.8 m s^{-1}

The height reached by the ash column can be estimated by different means. The formula for height in terms of mass erupted, M,

$$Ht = 236.6 \, M^{1/4} \tag{4}$$

gives 18.6 to 21.6 km, whereas in terms of the rate of release of thermal energy, Q,

$$Ht = 8.2 \, Q^{1/4} \tag{5}$$

gives 20.2 to 24 km. Q is the rate of release of thermal energy, 0.4×10^{18} J in 1.5 to 3 hours or 3.7 to 7.4×10^{13} W.

A study of grain sizes and their distribution showed that fragments measuring 3 cm fell as far as 7 km from the crater along the axis of the dominant wind direction and 5 km in a direction perpendicular to that axis. This indicates a column 21 km high and a tradewind of 10 m per second from the northeast. These results are consistent with an observed average column height of 20 to 22 km.

The relationship of the total height of the column (H_t) to that of the convective zone (H_b) and to the radius of the vent (r) shown in Fig. 7.6 is expressed by the formula:

$$H_t = 1.32 \, (8r + H_b) \tag{6}$$

A total height, H_t, of 20 to 22 km corresponds to a value for H_b of 14.7 to 16.2 km and a vent radius of about 0.06 km. The radius of the column reached 7 to 8 km.

The initial ejection velocity, U_o, of particles at the level of the crater can be estimated from the equation:

$$U_o = (8gr_o d/3Cd_g)^{1/2} \tag{7}$$

where g is the acceleration of gravity (9.81 m s^{-2}), r_o the size of the largest pumice fragments (0.4 m), d its density (a mass per unit volume of 1000 kg m^{-3}), C the coefficient of friction (approximately 1), and d_g the effective density of the gas in the crater (0.19 kg m^{-3}). Thus, U_o would be about 230 m s^{-1} for a fragment of 40 cm and 310 m s^{-1} for one of 70 cm—almost the speed of sound, During some eruptions of this kind, the velocities of fragments are probably supersonic.

All of these calculations are carried out after the fact on an event for which there are no historic records. At the time of the eruption (around AD 1300), the island of Martinique was inhabited by Carib Indians who left no account of the event.

Studies of the AD 79 eruption of Vesuvius, for which we have the classic descriptions of Pliny the Younger, enable us to retrace the sequence of events more precisely. It began on the 24th of August at one in the morning with two eruptive phases (Fig. 7.7). During the first 7 hours, the plume of ash rose 14 to 26 km with an output of 7.7×10^7 kg s^{-1} and a heavy fall of white pumice. This phase resembled the P1 episode of Mt. Pelée. Then, for the following 12 hours, the plume rose to a height of 32 km and the white pumice gave way to grey, indicating a chemical change in the composition of the magma. The pumice completely covered the city of Pompeii 9 km southeast of the volcano. A half-dozen violent pyroclastic flows, often referred to as "surges," were emitted (see Chapter 8). The first, on the 25th of August

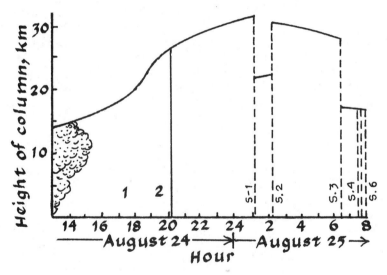

Fig. 7.7 Evolution of the height of the eruptive column during the course of the AD 79 eruption of Vesuvius. The time scale based on the account of Pliny the Younger assumes a constant rate of accumulation of pumice. 1 indicates the time when the roofs of buildings began to collapse in Pompeii. 2 marks the change from white to grey pumice. S1 to S6 are the especially strong surges and pyroclastic flows. (S1 covered Herculaneum, S2 reached Terzigno, S5 covered Pompei, and S6 reached Capo Miseno.) (Modified from Carey S., and H. Sigurdsson. 1987. *Geol Soc Am Bull* 99:303–14.)

at one in the morning overwhelmed Herculaneum, and the fifth, about 6 hours later, descended on Pompeii. In the course of the 19-hour eruption, 4 km³ (DRE) of tephra were emitted.

Eruptive columns produced by these large-scale explosive eruptions can enter the lower part of the stratosphere (at heights of 10 to 17 km). Those of Mt. Pelée and Vesuvius, which are estimated to have risen to 20 to 30 km, were exceeded by those of the eruption of Pinatubo, Philippines, in 1991 (40 km) and the prehistoric eruption of Taupo, New Zealand, where the columns rose more than 50 km. When ash from these eruptive columns is dispersed into the upper stratosphere, it can affect the climate on a global scale (see Chapter 14). The ash eventually falls over much of the earth. Although the fine, widely dispersed ash may not normally be noticed, it is preserved in the polar ice sheets, either as fine dust or as a weakly acidic layer in the ice.

7.5 ▲ Widespread Distribution of Ash

The distribution of airborne ash depends largely on the strength and direction of winds at different altitudes. Because high-altitude winds, mainly between about 5,000 and 15,000 m, trend more or less east–west, the distribution of recent ash is predominantly east or west of the eruptive vents, particularly in the tropics.

This wind effect has been observed in several historic eruptions. During the eruptions of Krakatau in 1883, fine ash rose to heights of more than 50 km; the ash fall was visually apparent 2,500 km west of the volcano, and the total area covered by easily recognizable ash was 827,000 km². Much impalpable dust remained in the upper atmosphere for several years, causing brilliant sunsets throughout the world. The dust clouds encircled the globe in 13½ days, and at an altitude of 30 to 50 km, their average velocity was 120 km per hour.

The spread of ash following the eruption of Mt. St. Helens on May 18, 1980, is well documented, for it was followed closely not only on the ground but also by continual satellite observations. In the first half hour, a vertical plume reached a height of 27 km and then formed a steady column between 14 and 19 km high. Within 10 minutes, a mushroom-shaped cloud had formed, and the prevailing winds began to carry ash hundreds of kilometers to the east (Fig. 7.8). The ash formed a layer 4 cm thick at 300 km from the volcano and was dropped in smaller amounts at distances of more than 1,500 km. Satellite photographs recorded the spread of the ash cloud across the states of Washington, Idaho, and parts of Oregon, Montana, and Wyoming. In 10 hours it traveled about 1,000 km for an average speed of 100 km per hour.

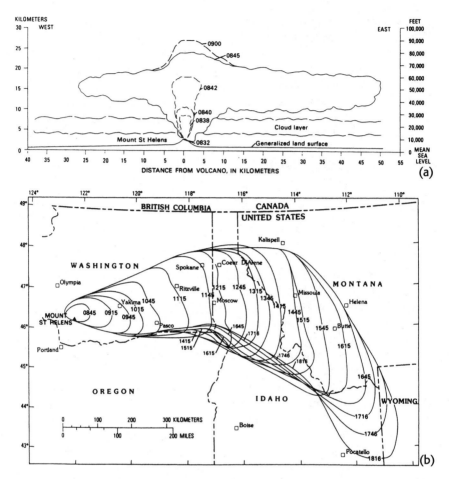

Fig. 7.8 The spread of the plume of ash and fine scoria following the eruption of Mt. St. Helens, Washington, on May 18, 1980. These extraordinary records were obtained by NOAA satellite images every half hour. **(a)** The ash plume reached an elevation of 27 km in half an hour and then spread into a mushroom form. **(b)** After 10 hours the ash cloud had crossed parts of five states. Under the influence of the prevailing winds, it drifted 1,000 kilometers to the east at an average velocity of 100 km per hour. (Sarna-Wojcicki et al., 1981. *US Geol Surv Prof Paper* 1250:577–600.)

Although the eruption on the 18th of May produced 1.1 km³ of ash, this was not the strongest explosive eruption in the history of the volcano. The stratigraphic record of well-dated ash deposits laid down by prior eruptions shows that Mt. St. Helens has been intermittently active for the past 4,500 years and that the magnitudes of many earlier eruptions have exceeded that of 1980.

7.6 ▲ Thickness, Fragmentation, and Dispersion of Deposits

Studies of tephra deposits can be used to quantify the intensity of explosive eruptions, even long after they have occurred. Three factors—thickness, grain size, and aerial extent—reflect the simple principle that violent eruptions produce ejecta of finer sizes and distribute them in greater thicknesses over wider regions. An example of the use of these factors in evaluating the prehistoric P1 ash fall of Mt. Pelée is discussed in an earlier section (Fig. 7.9). The different isopachs (lines

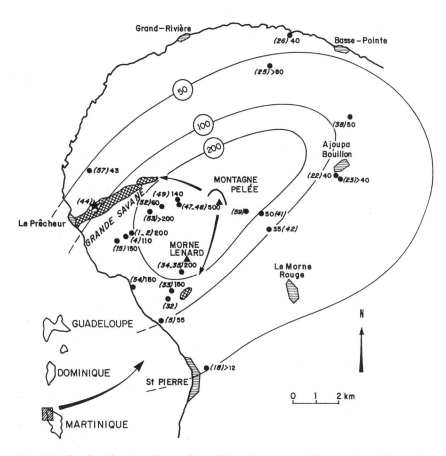

Fig. 7.9 The distribution of sampling of the P1 pumice-fall deposits of Mt. Pelée. Each sampling point is indicated by a dot with the number of the sample in brackets and thickness in centimeters. The isopachs are drawn as solid lines. The cross-hatched area indicates pumice flows associated with the same eruption, and the arrows show the directions in which they moved. (Bardintzeff, J. M., *et al.* 1989. *J Volc Geoth Res* 38:35–48.)

of equal thickness of a deposit) have an elliptical form that testifies to two regimes in the atmosphere above the vent: the southwesterly tradewinds at elevations of less than 6 km and winds in the opposite direction at higher altitudes.

Thickness (T)

The thickness of the deposit is a direct function of the total volume erupted and diminishes exponentially with distance from the vent. The thickness of pumice from the P1 eruption decreases from between 5 and 8 m at a distance of 2 km from the crater to only 0.4 m at a distance of 8 km. The factor T_{max} corresponding to the maximum thickness at the crater is difficult to measure because many fragments fall back into the crater and become reduced in size by subsequent eruptions. A graphical extrapolation of T_{max}, based on the variations of thickness with distance from the crater, gives a theoretical value of 15 m.

Fragmentation (F)

The degree of fragmentation, which increases with the strength of the eruption, corresponds to the percentage of material with sizes smaller than 1 mm. One uses the value at the distance where the isopach for 0.1 T_{max} (1.5 m in the case of the P1 pumice) intersects the main axis of dispersal, which, for this deposit, is NE–SW and corresponds to 10%.

Dispersion (D)

The aerial distribution of a deposit of this kind is difficult to measure because there is no easy way to determine the areas that were covered by only a few millimeters of ash and that were later swept clean by wind and rain. For this reason, the area enclosed by a well-defined isopach is normally used. The isopachs for 2, 1, and 0.5 m (Fig. 7.9) enclose areas of 26, 69, and 181 km^2, respectively. By extrapolation, one can calculate that the isopach for 0.01 T_{max} enclosed 900 km^2, and it is this value that is used for D.

Thus, the P1 eruption is characterized by the factors T_{max} = 15m; F = 10%, and D = 900 km^2. Although this method may seem a bit complex, it permits a quantitative comparison of various types of pyroclastic eruptions (Fig. 7.10). By way of comparison, the Hawaiian type has a very restricted aerial extent (D) and little or no fragmentation (F), whereas for the Strombolian type, these values are much larger. Two trends can be distinguished. One includes sub-Plinian, Plinian,

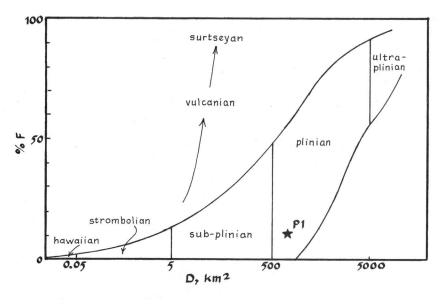

Fig. 7.10 Diagram for the factors fragmentation (F) and dispersion (D) of ash-fall pyroclastic deposits. The different types of eruptions are indicated, including the Hawaiian type in which fragmentation and dispersion are both very limited. The P1 eruption of Mt. Pelée falls within the field of Plinian eruptions. The prehistoric eruption of Taupo Caldera in New Zealand would be an example of an ultra-Plinian eruption. (From Walker, G. P. L. 1973. *Geol Rundschau* 62:431–36.)

and ultra-Plinian eruptions that last for hours and spread pumice over vast areas; the area within the isopach for 0.01 T_{max} may be as much as 50,000 km². The second trend includes violent Vulcanian and Surtseyian eruptions that produce very fine ejecta. The proportion of material deposited outside the 1-cm isopach is usually only 1 to 10%, but in exceptional cases it may be as great as 74%, as it was, for example, in the 1902 eruption of the Guatemalan volcano Santa Maria.

The forms that isopachs take are highly varied. Their simplest form is circular, but most are distorted by the effects of prevailing winds. The pumice deposits from the eruption of Mt. Mazama that formed Crater Lake are a good example (Fig. 7.11). They have two distinct lobes produced by two eruptions that were very close in time but were affected by different wind patterns. The form of isopachs for ancient eruptions may be difficult to establish; the products of one eruption are often mixed with those of another, and material may have been redistributed by erosion and redeposition. Nevertheless, the method can enable one to identify an eruptive center within a set of concentric isopachs, even long after the topographic form of the volcano is gone.

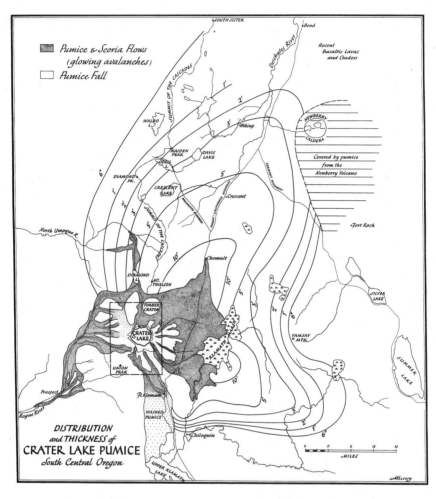

Fig. 7.11 A good example of a complex set of isopachs is that of the pumice-fall deposits from the eruption of Mt. Mazama that form Crater Lake. Thicknesses are shown in feet and inches. The prevailing wind was toward the north but an east-trending lobe was formed during a relatively brief shift of wind. (Williams, H. 1942. *Carnegie Inst Wash Pub 540*, 162 p.)

Where ash has been carried over the sea, a great many cores must be obtained from oceanic sediments to establish the pattern of distribution. For example, large numbers of undisturbed samples of ash recovered from the floor of the Mediterranean Sea have been used to define the distribution and volume of ash-fall deposits from the prehistoric eruption of Santorini (Fig. 7.12). In a similar way, particles

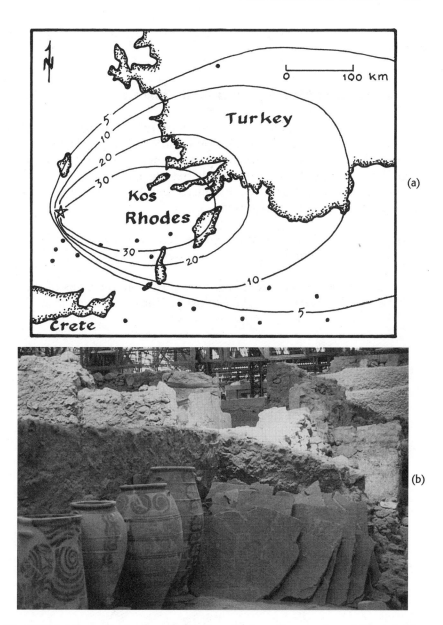

Fig. 7.12 (a) Isopachs for the ash-fall deposits from the eruption of Santorini in the 16-17th century BC have been compiled from deposits preserved on land and cores of sediments obtained from the seafloor. The data have been compiled from various sources by Pyle (1990). **(b)** Archeological excavations at Akrotiri have revealed a Minoan town buried under pumice. (**b,** Photograph by J. M. Bardintzeff.)

of ash have been recovered from peat deposits and glaciers where they were buried and preserved.

7.7 ▲ Glass, Lithic Fragments, and Crystals

The nature of the material constituting a particular pyroclastic unit is an equally useful tool. Although lapilli and scoriaceous Strombolian bombs may be very similar from one eruption to another, the ejecta of large Plinian ash-falls are distinctively heterogeneous.

Three genetically different components can be distinguished in the products of most Plinian eruptions.

1. The juvenile component is represented by vesiculated and angular fragments of fresh glass, often in the form of pumice.
2. Lithic fragments of older rocks are torn from the walls of the vent and pulverized by the explosion. They may come from

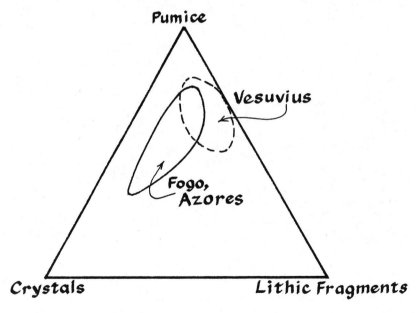

Fig. 7.13 Plinian eruptions: one from Fogo, Azores (46,000 ago) and the other from Vesuvius (AD 79). (Adapted from Walker G. P. L. 1981. *Bull Volc* 44:223–40.)

lavas of the same volcano or from older rocks of the underlying crust or mantle.

3. Crystals formed in the magma chamber or vent are often separated from the rest of the magma before or during the eruption. Their euhedral form distinguishes them from minerals derived from disintegrated older rocks.

The proportions of these three components are often very characteristic of an individual eruption and can be a useful tool in correlating eruptive units (Fig. 7.13). For example, when the ejecta of Mt. Pelée are separated into these three components, their size distributions can be classified according to the typical unimodal pattern for pyroclastic-fall deposits. Pumice is the dominant component among fragments of large size (>1 mm), whereas crystals make up the largest proportion of smaller particles. As we shall see in the chapter that follows, the size distribution makes it possible to distinguish the deposits from those of pyroclastic flows that have marked many of the same eruptions of that volcano.

7.8 ▲ Explosivity

A volcanic explosivity index (VEI) has been devised to provide a measure of the strengths of pyroclastic eruptions. Based on several numerical factors (Fig. 7.14), it resembles in a crude way the magnitude of earthquakes defined by the Richter scale and provides a semiquantitative means by which one can assess incomplete records of ancient eruptions. A ranking of seven degrees (0 through 6) is based on combinations of different criteria ranging from well-known facts to hypothetical inferences. These factors include volume erupted, duration, height of the eruptive column, and amount of ash reaching the troposphere or stratosphere.

Of a total of more than 8,000 recorded eruptions, about 20 emitted volumes in excess of 1 km³. Eleven Plinian eruptions, mostly prehistoric, produced between 10 and 25 km³. The eruption of Tambora, Indonesia, in 1815 produced 175 km³. In an average decade there are 104 eruptions of magnitude 0, 110 of magnitude 1, 304 of magnitude 2, and 82 with a magnitude equal to or greater than 3. In the last five centuries, there have been only four eruptions of magnitude 6 (Long Island in New Guinea in the 18th century; Krakatau, Indonesia, in 1883; Santa Maria, Guatemala, in 1902; and Katmai, Alaska, in 1912) and only a single eruption of magnitude 7, that of Tambora in 1815.

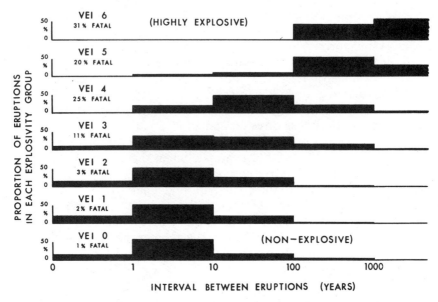

Fig. 7.14 Explosivity and time intervals between eruptions. The volcanic explosivity index (VEI) was devised by Newhall and Self (1982) to provide a semiquantitative measure of the magnitude of explosive eruptions. It is based on several factors, including the height of the eruption column, volume of ejecta, distance to which ejecta of a given size were thrown, and the number of casualties. Note that large eruptions tend to follow long periods of repose. (After Simkin, T., and L. Siebert. 1994. *Volcanoes of the world* [2 ed]. Missoula, MN: Smithsonian Institute, Geoscience Press.)

Suggested Reading

Cas, R. A. F., and J. V. Wright. 1988. *Volcanic successions, modern and ancient*. London: Chapman & Hall.
An excellent study of volcanism based chiefly on field relations.

Fisher, R. V., and H. U. Schminke. 1984. *Pyroclastic rocks*. Berlin: Springer-Verlag, 472 p.
An outstanding work on all aspects of the products of explosive eruptions.

Freundt, A., and M. Rosi, eds. 1998. *From magma to tephra*. New York: Elsevier, 318 p.
A collection of up-to-date papers dealing with the mechanics of pyroclastic eruptions.

Lipman, P. W., and D. R. Mullineaux, eds. 1981. The 1980 eruptions of Mount St. Helens. *US Geol Surv Prof Paper* 1250.
A collection of papers documenting all aspects of one of the most important eruptions of the 20th century.

Sparks, R. S. J., M. I. Bursik, S. N. Carey, J. S. Gilbert, L. S. Glaze, H. Sigurdsson, and A. W. Woods. 1997. *Volcanic plumes*. Chichester, United Kingdom: Wiley.
An authoritative study of the physical nature of large volcanic eruption columns.

Walker, G. P. L. 1971. Grain size charactersitics of pyroclastic deposits. *J Geol* 79:696–714.
An important reference on the principal types of tephra.

Walker, G. P. L. 1973. Explosive volcanic eruptions: A new classification scheme. *Geol Rundschau* 62:431–46.
A widely accepted method of classifing eruptions on the basis of observable quantitative factors.

Wilson, L. 1976. Explosive volcanic eruptions. III: Plinian eruption columns. *Geophys J Roy Astron Soc* 45:543–56.
A quantitative analysis of powerful explosive eruptions.

Wilson, L. 1980. Relationship between pressure, volatile content and ejecta velocity in three types of volcanic explosions. *J Volc Geoth Res* 8:297–313.
A study of the mechanisms of pyroclastic eruptions.

Pyroclastic Flows
and Lahars

▼

Pyroclastic flows are suspensions of hot pyroclasts, gas, and lithic fragments that are propelled across the ground surface, either by gravity or by directed explosions. Having high velocities and temperatures of several hundred degrees, these "pyroclastic hurricanes" can devastate the terrain for distances of tens of kilometers, even over varied topography. Very large pyroclastic flows, known as *ignimbrites*, can have enormous volumes and cover areas of thousands of square kilometers. Fortunately, the phenomenon is relatively rare. Although large ignimbrites have erupted many times in the recent geological past, none has been witnessed in historical times. On average, weak pyroclastic flows occur every year or so, but because they can be so devastating, they are a major hazard.

Lahars are also flows of fragmental debris, but they have lower temperatures and usually contain large proportions of water. More common than pyroclastic flows, they take an even greater toll on life and property.

Of all the types of pyroclastic flows, *nuées ardentes* are by far the most common. The name is often used interchangeably with the English term *glowing avalanche*. The name "nuée ardente" was coined in 1873 by Ferdinand Fouqué, who took it from *ardente nuvem*, the words used by the people of the Azores to describe the eruptions of San Jorge in 1580 and 1808. Fouqué's son-in-law, Alfred Lacroix, made the term famous when he used it to describe the 1902 eruption of Mt. Pelée that wiped out the city of Saint Pierre and its 28,000 inhabitants. Thanks to Lacroix's careful eyewitness accounts, the activity of Mt. Pelée in 1902 and 1903 has long been the type–example of this phenomenon.

8.1 ▲ The Nuées Ardentes of Mt. Pelée

For centuries, the activity of Mt. Pelée has been characterized by alternating pyroclastic falls, such as the one described in Chapter 7, and pyroclastic flows like those of 1902. The volcano has been especially active during the last two centuries. Before the terrible events of 1902, an eruption in 1851 produced moderate amounts of ash, and a weak steam eruption occurred in 1889.

Starting in February 1902, an odor of sulfur was noticed at Saint Pierre, a prosperous port situated at the foot of the volcano. During the last part of April, earthquakes and underground rumbling became common, and a heavy blanket of ash began to fall. On the morning of May 3rd, the ash fall was so heavy that it brought almost total darkness. During the nights of May 4th and 5th, a storm broke out, and lightening streaked across the sky. Thunder was heard, and streams draining the slopes of the volcano overflowed their banks. At 12:45 PM on May 5th, a mudflow swept down the valley destroying a rum factory and taking 25 lives. New magma became visible on the 6th of May when a small dome appeared in the bottom of the crater and a glow was seen during the night of May 7th. On the morning of Tuesday, the 8th of May, as church bells were ringing for the Day of Ascension, the volcano suddenly released its fury. At 8:01 a glowing avalanche descended the valley of the Rivière Blanche at 140 m per second (500 km hr^{-1}), swept through the city like a hurricane, and incinerated everything in its path. The supersonic shock wave (450 m sec^{-1}) preceding the main mass of debris caused the atmospheric pressure to surge by 36 mm of mercury in a few milliseconds. Twenty-eight thousand inhabitants died instantly, leaving only two survivors, a shoemaker named Léandre and a prisoner in the city jail named Cyparis. An area of 58 km^2 was completely devastated, and ships in the harbor were burned and capsized. The eruption was followed by others of declining intensity, notably on May 20th and 26th, June 6th, and July 9th.

Another eruption on the 30th of August was even stronger than that on the 8th of May. The nuées ardentes devastated an area twice as large (114 km^2 plus another 48 km^2 blanketed with ash) and caused an additional thousand deaths at the town of Morne Rouge. The eruptions continued for more than a year and discharged a total of about 60 pyroclastic flows of differing intensities during the rest of 1902 and early 1903.

During November 1902, a viscous dome and spine of light-colored dacite began to grow in the summit crater (see Chapter 6). The eruptions after the emergence of this dome differed from earlier ones in that they did not come from an open crater but from the base of the growing spine, often causing it to crumble.

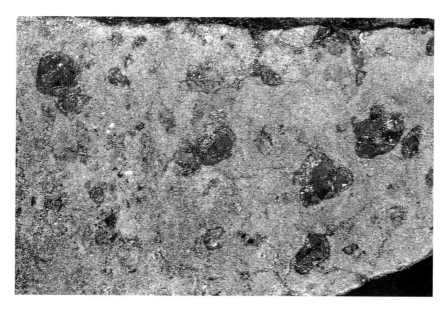

Fig. 1 A coarse-grained garnet lherzolite consisting of olivine (yellow), clinopyroxene (green), orthopyroxene (gray), garnet (red), and magnesian mica (brown) contains all the components, which, on partial melting, can yield a liquid of basaltic composition. Width of specimen is 4 cm. (Photograph courtesy of H. H. Schmitt and H. S. Yoder, Jr.)

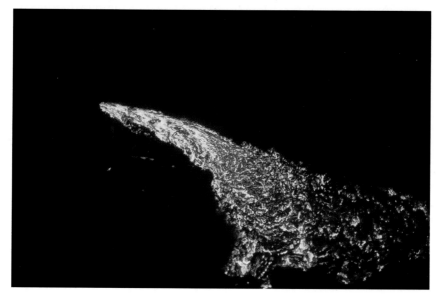

Fig. 2 A lava flow from Piton de la Fournaise, Reunion Island, in April 1998 is typical of the fluid basaltic lavas of oceanic volcanoes. (Photograph by J-M Bardintzeff.)

Fig. 3 A lava flow from Kilauea Volcano, Hawaii, in March of 1993 had a temperature in excess of 1100°C. (Photograph by J-M Bardintzeff.)

Fig. 4 A closer view of the same flow as Fig. 3. The low viscosity and high temperature are characteristic of pahoehoe lavas. (Photograph by J-M Bardintzeff.)

Fig. 5 A lava flow passing through a forest during the eruption of Mt. Cameroon in April 1999. (Photograph by J-M Bardintzeff.)

Fig. 6 A lava from Kilauea entering the sea in March 1993. Note the small amount of steam generated by the incandescent lava. (Photograph by J-M Bardintzeff.)

Fig. 7 The lava lake of Pu'U O'o on Kilauea in March 1993 had a temperature in excess of 1150°C. (Photograph by J-M Bardintzeff.)

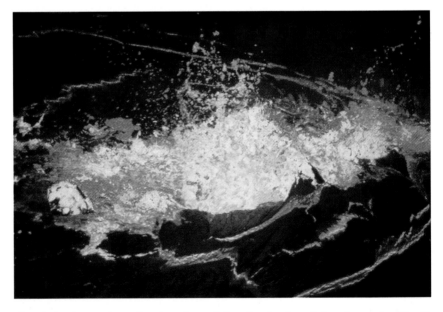

Fig. 8 A closer view showing fountaining in the lava lake shown in Fig. 7. (Photograph by J-M Bardintzeff.)

Fig. 9 Strombolian activity is defined on the basis of eruptions such as this which occurred in April 1978 at the volcano Stomboli in the Eolian Sea off the coast of Italy. (Photograph by J-M Bardintzeff.)

Fig. 10 A helicopter view of fountaining and a lava lake, 50 m in diameter, at Piton Kapor on the large volcano, Piton de la Fournaise, Reunion Island, April 1998. (Photograph by J-M Bardintzeff.)

Fig. 11 An explosive eruption from Tungurahua Volcano, Ecuador, on November 1, 1999, produced a plume of ash and gas 4 to 5 km high. (Photograph by J-M Bardintzeff.)

Fig. 12 A pyroclastic flow from Mount St. Helens on August 7, 1980. (Photograph by Robert Hoblitt, U.S. Geological Survey.)

Fig. 13 Mount St. Helens in August 1987. A small dome was growing in the horse-shoe-shaped crater. In the foreground are tree trunks swept into Spirit Lake by the 1980 pyroclastic flows. (Photograph by J-M Bardintzeff.)

Fig. 14 On the 8th of May 1902, the town of St. Pierre, Martinique, was destroyed by pyroclastic flows descending from Mount Pelée. (Photograph by J-M Bardintzeff.)

Fig. 15 Jacques-Marie Bardintzeff examining sulfur-rich deposits in fumaroles on Vulcano, one of the Eolian Islands near the coast of Italy. (Photograph by I. Bardintzeff.)

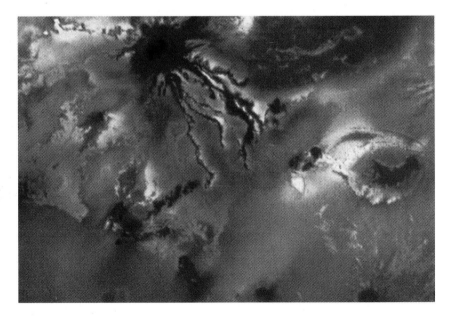

Fig. 16 Volcanism on Io, a satellite of Jupiter as seen from the spacecraft Voyager 1. The lava flows are 150 to 200 km long. (NASA photograph.)

Mt. Pelée returned to activity in 1929 and continued for another three years. During that period the American volcanologist, Frank Perret, made detailed first-hand observations of the eruptions, but the activity was less intense and caused no further loss of life.

The invaluable observations recorded by Lacroix and Perret showed how the pyroclastic flows were able to travel with such great velocities. Lacroix spoke of them as having two parts: a dense basal flow and an overlying ash cloud that largely concealed the basal layer (Fig. 8.1). The mobility of flows was mainly a function of the amount and temperature of the gas that kept the basal layer in suspension. Perret observed that the gas initially contained in the flow at its source was augmented by air picked up and heated when the flow passed over uneven ground. The contribution of small amounts of gas liberated from fragments in the moving flow was relatively minor.

More recent studies have shown that the large eruptions left two kinds of deposits: one of coarse, unsorted debris that followed the Rivière Blanche and another consisting of ash laid down in layers of varied thickness over a wide region. The first were deposited from the basal part of the flow, whereas the second type settled out of the

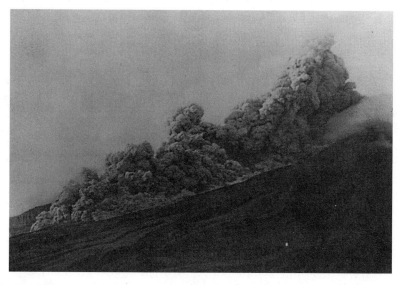

Fig. 8.1 This nuée ardente descending the slope of Mt. Pelée on January 25, 1903, resembles the one that wiped out the city of St. Pierre and its 28,000 inhabitants. Although the dark, turbulent cloud rising from the flow is very impressive, the most destructive part of the pyroclastic flow was the hot, debris-charged basal section seen at the leading edge of this example. (Photograph from A. Lacroix, 1904, *La Montaigne Pelée et ses Eruptions.* Paris: Masson, 662 p.)

expanded cloud of dust. Today only the distal parts of these deposits remain as the loose material left on slopes closer to the source has been stripped away by erosion. Partly for this reason, interpretations of the eruption on May 8th, based as they are on evidence found in the incomplete remaining deposits, are still subject to debate. Geologists disagree on whether there was a single pulse or a succession of weak, low-temperature pulses, followed by a more destructive glowing avalanche. The courses followed by individual phases of the eruption are too poorly known to answer this question.

After coming to rest, the deposits were degassed and compacted. The gray, heterogeneous dacitic material consists of blocks and lapilli in a matrix of ash containing large proportions of crystals. In contrast to airfall tephra, all these deposits are poorly sorted (Fig. 7.5a). Using a diagram showing the sorting coefficient as a function of median grain size (Fig. 7.5b), one can define the ranges of values that are characteristic of the two types of pyroclastic material deposited by falls and flows. Unfortunately, the two fields have considerable overlap. In some cases, size analyses of only the fraction smaller than 2 mm provide a clear distinction.

8.2 ▲ Velocities and Temperatures

An average nuée ardente from Mt. Pelée had a volume of several million cubic meters; it had about 10^{17} J of thermal energy and only about 10^{15} J of kinetic energy. The velocity was a function of the strength of the explosion, the mobility of the material, the slope angle of the volcano (between 10 and 15 degrees), the height of the source above the base of the volcano, and the distance from the source to the end of the flow. The distance to which a given pyroclastic flow can travel is delineated by an "energy line" that defines the limit that flows can reach over varied topography with the kinetic energy they gain on descending the flanks of the volcano. The relation between the horizontal distance and vertical drop serves as a useful basis for classification.

Velocities of relatively small pyroclastic flows (measured by timing how long it takes them to pass between two points a known distance apart) ranged between 17 and 50 m s^{-1} (60 and 180 km hr^{-1}) (Fig. 8.2). They vary greatly during the course of their flow. Lacroix observed that those of Mt. Pelée accelerated after leaving the dome and starting down the Rivière Blanche, then decelerated steadily after reaching the more gentle slopes at the base of the volcano. Some of the stronger flows continued out over the water of the bay.

Lacroix estimated the energy of the strong pyroclastic flows of Mt. Pelée by their ability to move large objects, such as the stone light

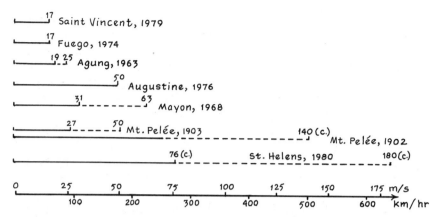

Fig. 8.2 Velocities of some typical nuées ardentes. The numbers give velocities in meters per second. The solid lines indicate average velocities, and dashed lines show maximum values. All are measured velocities except those marked *(c)*, which were calculated. (References are given in Bardintzeff, 1985a.)

tower at the entrance to the harbor. He calculated that the velocity needed to overturn the tower, which had a mass of 290 tons, a height of 15 m, and a diameter of 4 m, was at least 131 m s^{-1}. Similarly, a large statue of the Virgin Mary weighing 4,300 kg and measuring 3.15 m in height was overturned, but its 42.5-ton base remained in place. Because the force was insufficient to move the base of the statue, the velocity of the flow was estimated to have been less than 156 m s^{-1}.

Velocities can also be estimated from the height of obstacles that a pyroclastic flow was able to surmount. The potential energy transformed to kinetic energy as the mass descends from its origin is reconverted to potential energy when it rises again. Taking into account losses to friction, the energy balance for the descending path is:

$$\tfrac{1}{2} mV^2 = ([100 - F]/100)\, mgh_1$$
kinetic energy gained = potential energy lost

and for the ascent:

$$mgh_2 = ([100 - F]/100) \cdot \tfrac{1}{2}mV^2$$
potential energy gained = kinetic energy lost

where m is mass, V velocity, h_1 the height descended, h_2 the height of ascent, and F the loss to friction (probably about 30 to 40% for pyroclastic flows). This reasoning indicates velocities of about 200 m s^{-1} for certain prehistoric pyroclastic flows.

The 1980 eruption of Mt. St. Helens provided unusual opportunities to measure velocities from photographs and seismic records. It is estimated that the blast produced by the explosion on the 18th of May had a maximum velocity of 180 m s^{-1} (650 km hr^{-1}) and an average velocity of 76 m s^{-1} (275 km hr^{-1}) over a distance of 5 km. Initial velocities of up to 170 m s^{-1} were observed when later eruptions issued from the growing dome.

The temperatures of emplacement of pyroclastic flows can be estimated in several ways. The best method is one based on the infrared spectrum of carbonized wood obtained from the interior of pyroclastic deposits where the wood did not have direct access to the atmosphere. In the absence of oxygen, heat causes a progressive decomposition of the wood. Above 200°C, dehydration leads to a decrease of the O—H ion in the absorption spectrum (at 3,300 to 3,400 cm^{-1}). Around 275 to 300°C, the breakdown of cellulose and lignins results in a decrease of C—H and C—O—C bonds. Thus, a scale can be established to determine temperatures between 200 and 500°C with an error of less than 50°C. The method uses the original and final composition of the wood and a reference infrared spectrum to obtain the original and present balance between the chemical components of the wood (about 50% carbon, 43% oxygen, 6% hydrogen, and 1% nitrogen) and its surroundings. The values obtained for a number of samples from the eruptions of Mt. Pelée range from 300 to 400°C.

Measurement of the magnetic properties of ejecta also give useful information. When a cooling rock falls below the Curie temperature of about 580°C the iron–oxide minerals take on a magnetic orientation parallel to the earth's field. Rock fragments that are below this temperature when they fall have random orientations, whereas blocks coming to rest while still very hot acquire a uniform orientation parallel to the earth's magnetic field.

8.3 ▲ Origins and Types of Pyroclastic Flows

Pyroclastic flows can be divided into three general types: the *Peléan* type, which is a directed explosion from the base of a dome or spine of viscous lava; the *St. Vincent* type, which is more widely dispersed and comes from an open crater; and the *Merapi* type, which is an avalanche from the oversteepened slope of a dome. They can also be classified according to the size of fragments and the proportions of juvenile material in their deposits. One can distinguish between true pyroclastic flows and avalanches by measuring the median size of the ash fraction, that is, particles smaller than 2 mm. Products of the former are finely pulverized by violent explosions and are carried for great horizontal distances, whereas those of the more common type of gravity-driven avalanches tend to be coarser and do not travel as far from their source. The shape factor can be used to distinguish shards of vesiculated glass derived from the juvenile magma of true pyroclastic flows from angular fragments of older rocks abraded by mechanical processes at low-temperatures or phreatic eruptions (see Chapter 9).

Using these criteria, one finds a wide range between pure avalanches and true pyroclastic flows (Fig. 8.3). Avalanches differ in the amount of molten material involved, but in all cases, the magma is

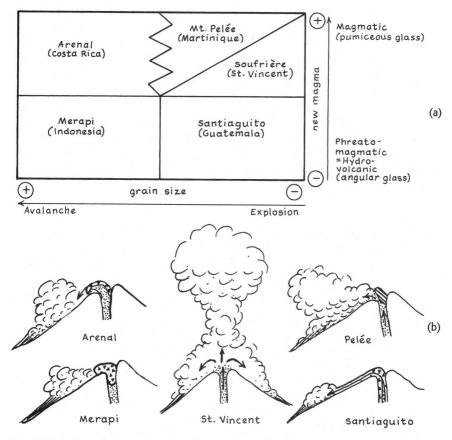

Fig. 8.3 (a) Classification of nuées ardentes on the basis of grain size and the morphology of glass shards. **(b)** Different types of nuées ardentes: Pelée, Soufrière of St. Vincent, Merapi, Arenal, and Santiaguito. Hachures indicate a solid dome, triangles a blocky dome, and dots juvenile magma. (**a** and **b,** From Bardintzeff, J. M. 1985a. *Bull PIRPSEV-CNRS* 109b; Bardintzeff, J. M. 1985b. *J Geodynamics* 3:303–25.)

very viscous. The Indonesian volcano Merapi has frequent avalanches from a growing dome, but the weak explosions are mainly from heated meteoric water and produce angular fragments from brittle rock. This is in contrast to the Costa Rican volcano Arenal, where the avalanches contain fine pumice derived from the molten core of the dome. With increasing amounts of gas, avalanches of the latter type grade into true pyroclastic flows. Again, the gas may be largely meteoric, as it was in the 1973 eruptions of the Guatemalan volcano Santiaguito or mostly magmatic as in the 1915 eruption in the Valley of Ten Thousand Smokes, Alaska. Gas-rich eruptions may come from an open crater, as in the case of the Soufrière on the island of St. Vincent, or from a growing viscous dome or spine, as at Mt. Pelée. Conditions can, of course, change from one eruption to another. Mt. Pelée, for example, is best known for the eruptions that occurred while the crater was being filled by growing domes, but in the past it has produced many flows of the St. Vincent type.

All of these types can be equally dangerous, mainly because they are so sudden and unpredictable. During the series of eruptions of the Japanese volcano Mt. Unzen that began in May of 1991 (see Chapter 6), avalanches and pyroclastic flows from the growing dome followed such a regular pattern that observers were tempted to witness them at close range. Then without warning, a more violent event on the 3rd of June spread over a much wider area and killed 43 persons, including three volcanologists.

8.4 ▲ Surges, Directed Blasts, and Debris Flows

Other types of eruptions resemble glowing avalanches in that they transport debris in the form of low-angle explosions, known as surges and directed blasts. Large debris flows differ in that they are driven by gravity and rarely have high temperatures. Although their mechanisms of emplacement are similar and the terms are often used interchangeably, the term *blast* is more appropriate for a sudden explosive event, usually triggered by an avalanche, whereas a *surge* is a less dense, outward-directed, high-velocity flow, usually from the base of an eruption column.

Pyroclastic surges have such distinctive forms that some volcanologists consider them a third type of eruption intermediate between pyroclastic falls and pyroclastic flows (Fig. 8.4). Their most distinctive feature is the small amount of fragmental material that they carry. This gives them a lower density than other types of pyroclastic flows. Most of these hurricane-like clouds come from a collapsing eruption column. They cover the topography with deposits ranging in thickness from a few centimeters or less on flat, elevated topography to much

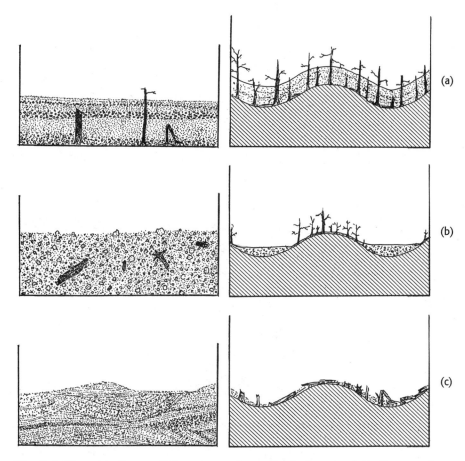

Fig. 8.4 Geometric florms of three types of pyroclastic deposits, falls, flows, and surges, laid down on the same type of topography. As shown in the left side of each figure, pyroclastic falls are usually bedded and well sorted, but deposits from pyroclastic flows are completely unsorted. Surge deposits are characterized by cross-bedding. As shown in the right side of each figure, deposits of pyroclastic falls mantle the topography with a thickness that is nearly independent of slope or elevation, whereas pyroclastic flows tend to follow valleys. Surges move across the topography scouring the surface on slopes facing the source (to the left in this illustration). Trees covered by pyroclastic falls may be stripped of leaves and small branches but normally remain standing. Trees knocked down by pyroclastic flows are commony charred and are randomly oriented in the deposits. Trees swept down by surges are aligned in uniform directions on slopes facing the source but are broken off or merely scorched where they are partly protected by topography.

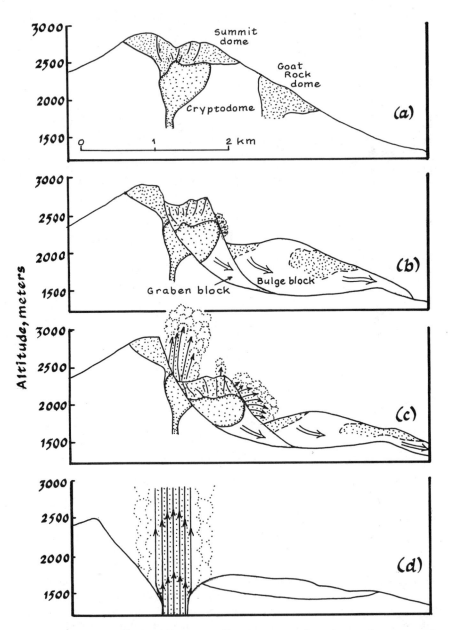

Fig. 8.5 Changes of the north–south profile of Mt. St. Helens during the course of the eruption on May 18, 1980. Based on interpretations of aerial photographs (Moore and Albee, 1981). **(a)** Before the eruption a cryptodome rose beneath an earlier dome formed in the 16th and 17th centuries. **(b)** The first explosions began 20 seconds after the gigantic landslide that uncovered the cryptodome. **(c)** After 30 seconds, sizeable explosions (directed blasts) took place from the blocks sliding down the flanks. **(d)** After the landslide had uncovered the main feeder, a new phase began with a large vertical eruption column. (From Moore, J. G., and W. C. Albee. 1981. *U. S. Geol Surv Prof Paper 1250*, 123–34.)

greater thicknesses in depressions. They have many of the characteristics of water-lain sediments, such as bedding, scour and fill, and cross stratification; individual layers are well sorted by size.

Directed blasts resemble surges but have a somewhat different origin and type of flow. The mechanism responsible for blasts has been identified from studies of the eruption of Mt. St. Helens on May 18, 1980 (Figs. 8.5 and 8.6), and that of Bezymianny in Kamchatka in

(a)

(b)

Fig. 8.6 (a) The 1980 blast from Mt. St. Helens flattened the forest over an area of 404 km². **(b)** Closer view of trees that were knocked down by the powerful blast. **(a,** Photograph by U.S. Geological Survey; **b,** photograph by A. R. McBirney.)

1956. In both cases, the avalanches and debris flows were produced by a slope failure that cut away an entire sector of the flank and summit of the volcano to a depth of several hundred meters. A shallow body of intrusive magma or "cryptodome" was suddenly exposed and the relief of pressure caused an explosive expansion of its compressed volatiles. Eruptions of this kind usually leave only a thin veneer a few centimeters thick, and although their volume is small they cover a wider region. (At Mt. St. Helens, for example, 0.19 km³ of such material were distributed over 600 km².) The proportion of juvenile material is greater than that of older rock, and large fragments are uncommon. Several explosions often follow in close succession before giving way to normal, gravity-driven pyroclastic flows.

Some directed explosions, such as that of Bezymianny, have involved large amounts of new magma, whereas others, such as that of Bandai-San, consisted solely of older rock. In both cases, a large, horse shoe-shaped crater was formed, but the explosive force of the Bezymianny type tends to be greater (VEI of 3 to 5 in contrast to 2 to 4 for those of the Bandai-San type), and the initial paroxysm is followed by growth of a dome or cone. Purely phreatic eruptions like that of Bandai-San rarely have subsequent activity other than fumarolic emissions.

Debris flows and avalanches consist mainly of older material with little if any magmatic component. They leave a large, amphitheater-like crater at their source and lay down deposits with a very distinctive appearance. In the case of Mt. St. Helens, 2.5 km³ were removed from the summit and spread over 64 km² to form a hummocky area with mounds of debris up to 400 m long, 150 m wide, and 20 m high. Deposits of chaotic debris, tens of meters deep, surround huge fragments of the mountainside that have remained intact. These large blocks, the so-called *block facies,* may be made up of any kind of older material, even ice, but most are hydrothermally altered rocks with little juvenile material from the dome or new magma. The finer material of the *matrix facies* consists of ash and pulverized rock fragments.

Among the avalanches of all sizes that have been recorded throughout the world, many have moved surprisingly large amounts of material. Individual Japanese volcanoes, for example, have produced anywhere from 0.03 to 9 km³ of avalanche debris, have lost between 200 and 2,400 m of their heights, and have left deposits extending for distances of as much as 32 km. Their small height-to-distance ratios of 0.2 to 0.06 testify to a great mobility, especially for very large avalanches. The avalanche deposits of Mt. St. Augustine in Alaska are good examples: those of Burr Point, which covered 20 km² in 1883, caused a tsunami when they entered the sea, and a blast from the same volcano around AD 1540 spread debris over 35 km².

8.5 ▲ Ignimbrites

A special term is needed for the huge pyroclastic flows that well up from fissures and spread with great mobility over wide regions. *Ignimbrites* are sometime called ash-flow tuffs, but most are much too coarse-grained to warrant that name. They differ from other pyroclastic deposits, not only in their mode of deposition but also in their aerial distribution, sorting, and degree of compaction and welding.

Most ignimbrites are products of large eruptions involving huge amounts of thermal energy. Although prehistoric ignimbrites have been identified in many regions, such as Nevada and Utah and large parts of Mexico and Honduras, no eruption of this kind has been observed directly. On average, they occur less often than once a century. The only ignimbrite eruption of this century occurred at the base of the Katmai volcano in Alaska, in what came to be known as the Valley of Ten Thousand Smokes. It began on the 6th of June in 1912 and lasted about 60 hours, but, so far as is known, there were no human witnesses. Large as it was, the volume of this ignimbrite, 7 km³, was small in comparison with those of many prehistoric eruptions, many of which are measured in thousands of cubic kilometers.

Geologists who first encountered these vast sheets were puzzled because they could not explain how pyroclastic rocks could be carried such distances through the air and still retain enough heat to be densely welded. The name *ignimbrite* literally means "incandescent rain," implying that they consisted of very hot pyroclastic-fall deposits. On closer examination, however, it was noted that unlike airfall tephra, which may bury trees but rarely burns or overturns them, ignimbrites contain logs and branches of all sizes reduced to charcoal and chaotically mixed throughout the deposit. Some geologists concluded from this that they were a strange kind of siliceous lava. Although some very hot airfall deposits may be welded if they accumulate rapidly close to their source, the term *ignimbrite* now connotes the deposit of a high-temperature pyroclastic flow that may or may not be welded.

Ignimbrites are erupted as hot, foaming, pumiceous magma that wells up from craters or fissures much in the manner that a warm carbonated beverage flows from a bottle when the cap is removed. The velocities with which they spread can be 200 m s⁻¹ (720 km hr⁻¹) or more, and some are capable of crossing obstacles as high as 700 m. They owe their great mobility to their low viscosity. Although the viscosity of the silicate melt alone is quite high—between 10^9 and 10^{14} Pa S—that of the suspension of particles in hot gas is many orders of magnitude less. Some of the gas is exsolved from the vesiculating magma, but most is probably trapped air that, on being heated, expands and keeps the particles in suspension.

Owing to their great mobility and ability to cover wide regions, ignimbrites can retain sufficient heat to be densely welded far from their source. When laid down in quick succession, they form a single *cooling unit,* commonly in excess of 100 m thick. Single cooling units in the San Juan Mountains of Colorado range from 1,300 to 1,800 m in thickness; before compaction, some must have been about 2,500 m thick. The ignimbrites around Taupo, New Zealand, cover 20,000 km², and some of the sheets around the Yellowstone calderas are almost as extensive. From Early Oligocene to Late Pliocene time, most of Nevada and much of western Utah were inundated by a thick succession of ignimbrites that covered more than 200,000 km² and had an aggregate volume of more than 50,000 km³. The thickness of individual sheets varies greatly, but most are meters to tens of meters thick.

Despite their great volume and extent, ignimbrites have a remarkably homogeneous appearance from one locality to another. They tend to have a relatively fine-grained matrix made up largely of juvenile material in the form of angular glassy shards of pumice in differing stages of compaction and welding. They grade from a dense, commonly welded base to a surface layer consisting of loose fragments of pumice.

Two factors affect the thickness of a cooling unit: the lateral extent and the degree of compaction and welding. Where confined to valleys, they tend to be much thicker. For example, the ratio of thickness to lateral extent is about 1 to 400 for the ignimbrites of the Valley of Ten Thousand Smokes but only 1 to 100,000 for the products of a more cataclysmic eruption that spread over flatter terrain around Taupo, New Zealand. In the first case, the deposits flowed down a valley, whereas in the second, they spread in all directions with little regard for topographic relief.

Some ignimbrites are welded, whereas others are not, regardless of their thickness. In the presence of water vapor, rhyolites become welded at temperatures as low as 535°C; pumice of other compositions or smaller water contents may require 600 to 750°C. Welding is most pronounced in ignimbrites containing few if any phenocrysts, probably because they were erupted at more elevated temperatures. Viscous magmas tend to be less strongly welded than more fluid ones. Strongly welded ignimbrites are similar in some ways to lava flows; they are dense and may even have columnar joints. Normally, ignimbrites come to rest and cool without further movement, but a few "rheo-ignimbrites" deposited on steep slopes may be remobilized under the force of gravity and take on the appearance of lava flows.

The discharge of these huge volumes of magma often leads to collapse of a caldera. As we shall see in Chapter 10, most caldera-forming ignimbrites are preceded by Plinian or ultra-Plinian eruptions. The airfall deposits from these opening phases form distinctive tephra layers underlying the ignimbrites (Fig. 8.7). The transition from one mode of eruption to another normally occurs when the eruption column collapses and the gas-rich magma radiates outward. At the same time, the rate of discharge increases as the vent becomes larger. It may also take place when the amount of exsolving volatiles decreases and the velocity of discharge declines (Fig. 8.8). Most ignimbrites that are preceded by a Plinian phase are emplaced at relatively low temperatures and are only weakly welded, indicating that these ignimbrites spread as pyroclastic flows derived from a less gas-rich part of the same magma as the preceding airfalls.

Fig. 8.7 Well-stratified tephra from the opening eruptions of Novarupta in 1912 are overlain by unsorted pumice flows of the Valley of Ten Thousand Smokes, Alaska. Note that limbs of trees were stripped bare by the pumice fall, but the trunks remained upright. This is in contrast to the pumice flow, which carried away the trees so that charred logs are chaotically mixed in the unsorted pumice. The figure of the geologist at left-center indicates the scale. (Photograph by B. B. Fulton in Fenner, 1920.)

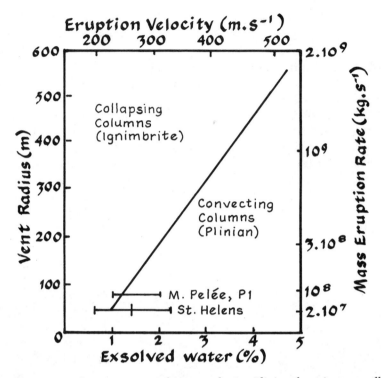

Fig. 8.8 Passage of a convecting column producing Plinian deposits to a collapsing column that produces ignimbrites. Values are shown for Mt. St. Helens (1980) and the P1 eruption of Mt. Pelée (circa AD 1300). (After Wilson, L., R. S. J. Sparks, and G. P. L. Walker. 1980. *Geophys J Roy Astron Soc* 63:117–48.)

8.6 ▲ Lahars

Pyroclastic flows are often difficult to distinguish in outcrops from a similar type of fragmental debris flow known as *lahars*. Both are characteristically unstratified and unsorted. A deposit containing blocks of varied compositions with rounded shapes is more likely to have been formed by a lahar than by a glowing avalanche. Most lahars become finer grained, better sorted, and more stratified away from their source, but this is rarely true of pyroclastic flows. Flattened pumice lumps, welding, and strong hydrothermal alteration are also diagnostic of hot pyroclastic flows.

Autobrecciated lavas may grade into lahars as they flow downslope, particularly if they enter water. Eocene lahars that originally covered almost 13,000 km² in the Absaroka Range of Wyoming were almost certainly formed in this way. The Mio-Pliocene Mehrten formation of the Sierra Nevada is similar; it once covered 30,000 km^2 and had a volume of about 8,000 km^3.

Lahars are not necessarily direct products of eruptions, even though they are indirectly of volcanic origin. Their name was coined in Indonesia, where they are especially common. They have also been referred to as "mudflows," but in recent years, this term has been shunned because it does not imply a volcanic origin. Any destabilizing event, such as an earthquake, landslide, or heavy rain, may mobilize ash or other fragmental debris that has accumulated on the flank of a volcano. The water they contain may have any of several possible origins; it may be volcanic, but more often it comes from rain, melted snow or ice, or from a crater lake. The fluidized mixture of water and rock fragments of all sizes can flow down slopes with great speed and destructive force. Temperatures are normally low unless the volcanic material has been erupted so recently that it still retains heat. Soil and clay produced by fumarolic alteration of older rocks eventually hardens into a cement-like material that binds the mass into solid rock.

Lahars are most common on high volcanoes capped by snow or ice or in tropical regions with heavy, prolonged rainfall. They can be found, for example, on the lower flanks of almost all the large andesitic cones of the High Cascades and Andes. Most of the city of Tacoma, Washington, is built on extensive lahars that came down the slopes of Mt. Rainier about 4,800 years ago. Records going back to 1539 show that at least 14 large lahars have descended more than 250 km from the volcano Cotopaxi in Ecuador, some with velocities of 80 km per hour. They were triggered by passage of glowing avalanches over glaciers around the summit of the volcano.

A dramatic example of the destructive power of lahars was seen on the 13th of November 1985 on the slopes of the Columbian volcano Nevado del Ruiz. A torrent of mud and debris flowed as far as 80 km and covered the town of Armero and other villages along with 25,000 of their inhabitants. This catastrophe is discussed in detail in a later chapter devoted to volcanic risks (Chapter 14).

For several years after the major 1991 eruption of Pinatubo in the Philippines, lahars continued to descend from the ash-covered slopes. The phenomenon was repeated during each rainy season and, if one judges from observations that followed smaller eruptions of Mt. St. Helens in 1980 and Galunggung in Indonesia in 1983, could continue with exponentially declining strength for another two decades. For the rainy season of 1991 alone, 200 lahars came down the flanks of the volcano, leaving deposits with a total volume of 0.38 km^3. Thicknesses range from 0.5 to 5m, but most deposits are between 1.5 and

Fig. 8.9 The town of Bacolor, situated 30 km from the vent of Pinatubo, was destroyed by lahars after the typhoon Mameng on October 1, 1995, four years after the eruptive phase of 1991. The thickness of the deposits at this locality was about 2 m. (Photograph by J. M. Bardintzeff.)

Fig. 8.10 Helicopter view of the routes of lahars from Pinatubo. (Photograph by J. M. Bardintzeff.)

3 m (Fig. 8.9). The principal rivers draining the volcano are the preferred routes of the lahars (Fig. 8.10), but when a valley is filled, new ones are incised in the easily eroded material. Those that obstructed the valley of Mapanuepe dammed a lake. With velocities of about 30 km per hour, the lahars flowed even on very gentle slopes to distances of several tens of kilometers. Many were still relatively hot when they came to rest; water percolating through the pumice-bearing material emerged at temperatures between 30 to 60°C.

As we shall see in the next chapter, lahars are particularly common after subglacial eruptions. Eruptions under ice produce large amounts of melt water, which mixes with fragmental volcanic debris to produce sudden floods that the Icelanders call "jökulhlaup."

Suggested Reading

Bardintzeff, J. M. 1985. Calc-alkaline nuées ardentes: A new classification. *J Geodynamics* 3:303–25.
A study of the principal types of a pyroclastic flows from andesitic and dacitic volcanoes.

Boudon, G., and A. Gourgaud, eds. 1989. Mount Pelée. *Jour Volc Geoth Res* 38: 200 p.
A collection of recent studies of one of the world's most renown volcanoes.

Fisher, R. V., and H. U. Schminke. 1984. *Pyroclastic rocks.* Berlin: Springer-Verlag, 472 p.
An excellent general reference on all types of pyroclastic eruptions and their products.

Lacroix, A. 1904. *La Montagne Pelée et ses éruptions.* Paris: Masson.
A classic work in which pyroclastic flows were first described in detail.

Moore, J. G. 1967. Base surge in recent volcanic eruptions. *Bull Volc* 30:337–60.
A lucid description of the eruptions in which pyroclastic surges were first recognized.

Newhall, C. G., and R. S. Punongbayan, eds. 1997. *Fire and mud: Eruptions and lahars of Mount Pinatubo, Philippines.* Seattle: University Washington Press, 1126 p.
A comprehensive account of the 1991 eruption of Pinatubo and of the devastating lahars that followed.

Walker, G. P. L. 1983. Ignimbrite types and ignimbrites problems. *J Volc Geoth Res* 17:65–88.
The classification and flow mechanisms of pyroclastic flows.

Wilson, L., R. S. J. Sparks, and G. P. L. Walker. 1980. Explosive volcanic eruptions. IV: The control of magma properties and conduit geometry on eruption column behavior. *Geoph J Roy Astro Soc* 63:117–48.
A thorough discussion of the mechanisms of pyroclastic eruptions.

Volcanism under Water and Ice

▼

By far, the greatest proportion of the earth's volcanism is under the sea, where the reaction of magma with water has a major effect on eruptive behavior. At shallow depths the marked increase of volume that accompanies the change from liquid water to steam can give eruptions a highly explosive character. The general term *hydrovolcanic* is used for all conditions in which the water is of external origin, but the influence of this water on the magma may be either direct or indirect. In purely *phreatic* eruptions, heated meteoric water expands explosively to steam, and little or no molten magma is erupted. In *phreatomagmatic* eruptions, the water comes into direct contact with the magma, and the two are erupted together.

9.1 ▲ Interaction with Water

If water is heated above its boiling temperature at low pressures where it can expand freely, the change entails a great increase of volume; however, if that expansion is inhibited, as it is in the pore–space of rigid rocks of low permeability, enormous pressures are generated. This combination of great pressure and expansion is responsible for the violent explosions produced by magma rising into shallow water or water-saturated sediments.

The interaction with water is much different when magma is erupted on the seafloor. Under these *supercritical* conditions (Fig. 9.1), the load pressure of overlying water limits the expansion of steam, and there is no sharp change in density, such as that between liquid and vapor at subcritical temperatures. Steam explosions are therefore suppressed, and the thermal energy is absorbed in heating a large mass of water.

143

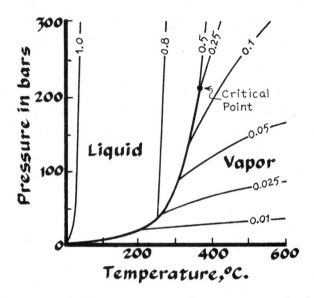

Fig. 9.1 The amount of volumetric expansion that accompanies the change from liquid to vapor depends on the pressure under which the transition takes place. As pressure increases, the boiling temperature also increases, but the amount of volumetric expansion declines. For example, an increase of pressure from one atmosphere to 100 bars raises the boiling temperature from 100 to 305°C and reduces the relative volumes of vapor to liquid from 1603 to about 165. At the *critical point* of 221 bars and 374°C, the density of the vapor is the same as that of the liquid, and there is no discontinuity between the two forms. Thus, at the pressures prevailing on the seafloor, about 500 bars, water has no well-defined boiling temperature, and the amount of expansion when water is heated by magma is greatly reduced.

Three factors govern how much of the thermal energy in magma is converted to mechanical energy: (1) the ratio of water to magma; (2) the physical state of the water, that is, whether it is a super-critical vapor or saturated steam (a mixture of liquid and vapor); and (3) the density of the mixture of solids and fluid. For a small water–magma ratio (<0.1), water plays only a minor role, and the magma can erupt in the form of lava or scoria. For ratios between 0.1 and 0.3, the greater proportion of thermal energy converted to mechanical energy in a limited mass of water causes explosive expansion and fragmentation of the erupting magma in the form of pyroclastic flows or phreatomagmatic surges. At larger ratios, the efficiency of energy conversion declines because the thermal energy is dissipated in a larger mass of water. Fragments produced during the interaction of magma and water have dimensions of less than about 50 microns for intermediate water–magma ratios and between 1 and 10 mm for smaller or larger ratios.

The rate at which heat is lost from the surface of lava is very sensitive to the texture of the surface, particularly the surface area. Even at shallow depths, smooth, pahoehoe or pillow lava can enter water quietly without producing large amounts of steam. A thin layer of steam insulates the lava from the main mass of water and retards the transfer of heat. Rough aa lava, because it has a much greater surface area, heats the water more effectively and generates more steam and turbulence. More viscous lavas and the aa varieties of basalt may become completely disrupted to form masses of unsorted fragments, sometimes referred to as *aquagene breccias* (Fig. 9.2). Brecciation is

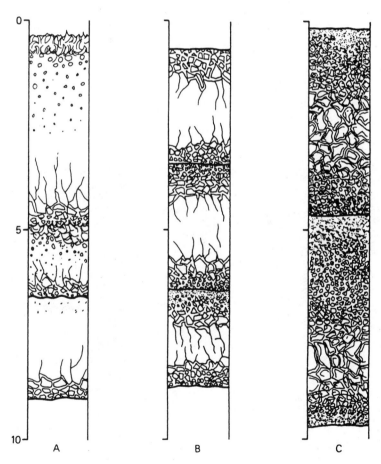

Fig. 9.2 Different types of submarine breccias from drill cores through atolls in the Tuamotu Archipelago. The scale is in meters. *A* and *B* are autobrecciated lavas; *C* is an autoclastic breccia. (Varet, J., and Demange, J. 1980. *Initial reports of the DSDP* XLIX:749–60.)

particularly prevalent in submarine lavas that have moved down a sloping surface.

9.2 ▲ Products of Hydrovolcanism

Eruptions through shallow water or wet sediments are the most familiar form of hydrovolcanism. Both phreatic and phreatomagmatic eruptions produce large amounts of debris in the form of angular lithic fragments of all sizes. The most distinctive products of such eruptions are *accretionary lapilli*. Sometimes referred to as *pisolites,* these are pellets with diameters ranging from about 1 mm up to 4 cm in diameter. They are formed by agglutination of wet ash, and although they tend to disaggregate easily, they are the most diagnostic products of this type of eruption. The magmatic component of phreatomagmatic eruptions usually takes the form of angular fragments of glass, but in the case of siliceous magmas, it may be pumice. Separate, well-formed phenocrysts of pyroxene or plagioclase may form crystal-rich deposits known as crystal–lapilli.

Fig. 9.3 Ejecta from Mt. Zini on the island of Kos (Greece). Very coarse avalanche deposits overlie finely stratified and cross-bedded hydrovolcanic tuffs. Cross-bedding such as this is characteristic of surge deposits (Photograph by H. Traineau.)

Larger types of ejecta may be either bombs or blocks. The former have a chilled outer layer of black glass with a surface that is mammillated like that of a head of cauliflower. When the interior continues to vesiculate, the expansion produces deep cracks in the rind. Many of these bombs break into wedge-shaped fragments, and their impact on soft, wet deposits produces indentations known as "bomb sags."

Much of the debris ejected during eruptions falls back into the crater, often through many cycles. In this way, fragments that were originally angular are abraded and rounded until they resemble water-worn sediments. The ejected debris from phreatomagmatic eruptions accumulates in poorly sorted deposits close to the vent, but in weaker eruptions, particularly when the proportion of magmatic material is small, successive explosive pulses lay down thin beds that resemble finely layered sand and silt (Fig. 9.3). Individual beds are graded with particle sizes decreasing from the base upward.

A distinctive type of ejecta, referred to as *hyaloclastite* (Fig. 9.4), is very characteristic of phreatomagmatic eruptions in which fragmentation results not only from explosive forces but also from thermal shattering of the quenched basaltic magma. Rich in glass and easily

Fig. 9.4 Hyaloclastite breccias were formed when the Columbia River basalts flowed into lakes and streams. In the example shown here, the lava was violently disrupted when it went into a river bed, but as the flow filled the channel and was able to flow without being submerged, it took on a more normal character. (Photograph by A. R. McBirney.)

altered, much of it becomes hydrated to a yellowish form known as *palagonite* with as much as 20% absorbed water. Large amounts of these hyaloclastites blanket the emergent oceanic ridges of Iceland and Ethiopia. In the latter locality they were erupted under water and later raised above sea level, but those of Iceland were formed by eruptions under the ice that covered most of the island during Pleistocene time.

9.3 ▲ Volcanism on the Deep Ocean Floor

Although eruptions on the seafloor are inconspicuous and seldom manifest at the surface, they constitute by far the greatest proportion of the earth's volcanism. This is especially true of the oceanic ridges where about 20 km³ of basalt are added to the crust each year, mainly in the form of dikes, sills, and lava flows erupted from fissures. Large volcanoes rarely develop along the ridge axes, but thousands are scattered over vast areas of the ocean floor.

Eruptions of a submarine volcano vary greatly between its birth on the deep seafloor and its emergence above sea level. At normal depths of 4 to 5 km, the pressures at the seafloor (about 500 bars) greatly restrict the expansion of exsolved gases or seawater coming in contact with the magma. Pyroclastic rocks are therefore rare until the volcano rises to shallower depths. Basalt, typically in the form of *pillow lavas* but also as intrusive sheets, is the most characteristic product of these volcanoes. Pillow lavas are less common on fast-spreading oceanic ridges, where slopes tend to be more gentle.

Pillow lavas are probably the most abundant volcanic rocks on earth. They are formed at all depths, even under shallow lakes and glaciers. Their formation has been observed at close range and even filmed by divers when a flow descended from Kilauea, Hawaii, and entered the sea (Fig. 9.5a and colored photo). Studies in Hawaii indicate that subaqueous flows develop a system of interconnected, cylindrical tubes that, when seen in cross section, have a more or less circular shape (Fig. 9.5b). This tubular structure is rather common on moderate slopes, but very fluid lava on steep slopes tends to form tongues or "toes," many of which become detached and form true spheroids.

Although the interaction of magma and seawater on the deep ocean floor is much less explosive, it can result in large hydrothermal systems, particularly in the vicinity of active ridges where seawater infiltrates down through fractures and returns at high temperatures. Close to the vents, the hot water, heavily charged with sulfur, deposits various compounds dissolved from the underlying basalts and sediments.

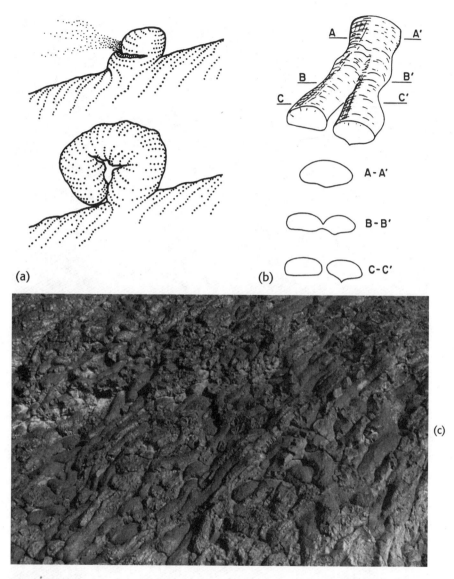

(a)

(b)

(c)

Fig. 9.5 (a) Formation of pillow lava by budding from a lava tube. The carapace cracks under the force of internal pressure causing an offshoot to form with a new crust around it. The new bud lengthens into a cylinder fed from the same or new fractures, and in some cases curves back on itself. **(b)** Pillow lava forming a system of branching tubes as seen in a series of cross sections. **(c)** Pillow lavas in submarine lavas now exposed on land in Oman. (**a,** From Moore, J. G., et al. 1973. *Geol Soc Am Bull* 84:537–46; **b,** from Vuagnat, M. 1975. *Bull Volc* 39:581–9; **c,** photograph by A. R. McBirney.)

continued

(d)

(e)

Fig. 9.5 continued (d) Pillow lavas exposed in a sea cliff on South Island, New Zealand. **(e)** Many pillows have an outer shell with radial joints and a massive interior that may or may not be vesicular depending on the depth of water. (**d** and **e,** Photographs by A. R. McBirney.)

These hydrothermal chimneys (Fig. 9.6), often referred to as *black smokers*, are rich in sulfides and have temperatures as high as 350 to 400°C. A second variety, known are *white smokers*, is rich in sulfates and has lower temperatures, usually in the range of 160 to 300°C. Both types represent substantial discharges of thermal energy. They have been found to support an abundance of marine life, much of it unique to this unusual environment.

Fig. 9.6 Strong hydrothermal activity in the vicinity of oceanic spreading ridges takes the form of "black smokers" such as this one near the Juan de Fuca Ridge. Note the dark, smoke-like column emerging from the chimney that has been built up over time by precipitates from the mineral-charged emanations. The heat and nutrients supplied by these vents support a wealth of strange marine organisms found nowhere else in the world. (Courtesy of E. A. Mathez of the American Museum of Natural History.)

The large amounts of heat released by these deep-seated manifestations can create huge thermal plumes that are capable of reaching the surface and altering the local temperature and circulation of the ocean.

9.4 ▲ Seamounts, Guyots, and Atolls

The ocean floor is littered with tens of thousands of submarine volcanoes or *seamounts*, many of which are vastly larger than any volcano on land. Little is known about the earliest stages of growth of volcanoes in the deep ocean basins where thick sediments blanket the sea floor. Because of the large density contrast between the basalt and wet, poorly consolidated silt and clay, it is unlikely that magma rises through any substantial thickness of light sediments to flow out on the surface unless a cone already protrudes through the sedimentary layer. In their youthful stages, volcanoes must grow beneath a cover of sediments and their own clastic debris much in the manner of Icelandic volcanoes that have grown under ice.

The small cone of Loihi near the southern base of the island of Hawaii is an example of an oceanic volcano in the early stages of growth. If it continues to grow, many centuries will pass before it reaches sea level. It has been studied closely for several years, during which time it has had at least one eruption. The lavas are extruded quietly with little or no visible manifestation at the surface of the sea. At the other extreme, Macdonald seamount, discovered by a bathymetric survey in 1967, exemplifies a volcano that is about to emerge above the water. When surveys were carried out between 1969 and 1975, its summit was situated at a depth of only 49 m. Measurements in 1981 and 1983, just after a period of activity that lasted from 1977 to 1981 showed that the summit plateau was at a depth of 40 m and a sharp central peak rose to within 27 m of the surface. A submarine eruption was actually observed in 1989 by the submersible vessel Cyana. The discolored water usually seen above the volcano indicates that its volcanism is associated with almost continual hydrothermal activity.

Evidence of deep submarine eruptions is rarely visible at the surface, because the steam they produce is quickly condensed as it rises through cold seawater, but if the eruption is vigorous and the vent is within 500 m or so of the surface, the water may become turbulent and cloudy. One of the best-documented examples was that of the early submarine eruptions of a new islet that appeared near the island of Iwo Jima near Japan in 1934. Activity was first detected when the vent was still at a depth of more than 300 m below sea level. Ash-laden steam reached the surface, but sulfur gases, which later proved to be an important volatile component, were removed by absorption in the seawater. During early phases of the eruption, large vesicular blocks with incandescent cores rose to the surface, floated until they cooled, and then sank. This was in contrast to other eruptions, such as one that broke through 30 m of water in the South Sandwich Islands in 1962. In the latter instance, large amounts of pumice reached the surface and continued to float long enough to be dispersed over a wide region. The large amount of inflation a magma must have to float in this way is possible only in very shallow eruptions; it is most common in siliceous material. Pumice is buoyant while the vesicles are filled with steam, but as the steam condenses the resulting vacuum draws in seawater and may cause the pumice to sink.

Shallow submarine eruptions can be spectacular (Fig. 7.1d). The name *Surtseyan*, used for eruptions of this kind, is taken from that of a new island that emerged on November 14, 1963, off the southern coast of Iceland. After the cone became large enough to exclude the seawater, the eruption took on a Strombolian character and lava armored the cone against the erosive action of waves. A similar eruption of the volcano Capelinhos began in September 1957 offshore from the island of Fayal in the Azores. Jets of steam during the opening phase were followed by incandescent scoria, which rose to heights of about 100 m. After 48 hours, a cinder cone 100 m high had risen above sea level, and by October, blocks and bombs were being thrown as high as 500 m. The accompanying ash plume rose to 7,000 m. At the end of the month, the new island had taken on the shape of a horseshoe when it suddenly stopped erupting, and part of the island was removed by wave erosion and underwater slumping.

When activity subsides and eventually comes to an end, erosion and subsidence begin to reduce the emergent parts of the volcano. Unless the surface is armored by resistant lava, waves rapidly cut into the cone. In addition, the volcano begins to subside slowly under its own weight. The enormous mass of large seamounts exert an irresistible load on the crust and underlying mantle. Lacking new lava to offset this subsidence, they begin to lose elevation. After subsiding below the surface, the truncated peaks of these flat-topped *guyots* testify to their having once been islands.

In tropical waters, corals colonize the offshore slopes down to depths of about 30 m. These fringing reefs form a segmented ring separated from the shore by a shallow lagoon. While the volcano remains emergent, the reefs are mixed with lava, tephra, and erosional debris, but as the edifice sinks, they continue to keep pace with subsidence and may eventually form a ring-island or *atoll*. Because coral grows only in warm water, atolls are not formed beyond a tropical belt extending about 28 degrees north and south of the equator in the western Pacific. If the drifting oceanic plate carries the volcano outside this limit, known as the Darwin Line, it enters cooler water, the coral

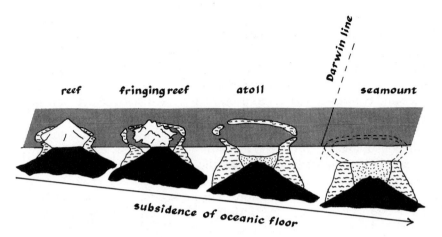

Fig. 9.7 Evolution of a volcanic island into an atoll and finally a guyot as the volcano subsides into the cooling oceanic lithosphere. When the structure crosses the Darwin line, coral can no longer grow and keep pace with subsidence of the volcano. The massive volcanic edifice also subsides by simply sinking under its own weight. (From Caron, J. M., et al. 1995. *Comprendre et enseigner la planète Terre*, 3rd ed. Paris: Ophrys.)

ceases to grow, and the atoll subsides below the water to become a coral-crowned guyot (Fig. 9.7).

A fossil reef only 13,900 years old has been found at a depth of 150 to 160 m in the Hawaiian chain; some of those in the western equatorial Pacific are even deeper. The combined effects of subsidence at a rate of about 2.5 cm per year and the rapid rise of sea level since the end of the last glacial period result in a net rate of submergence of about 10 cm per year, much more than the rate at which coral can grow. In the Polynesian chain, the average rate of growth of the volcanoes is estimated at 0.31 cm per year for Tahiti, and for Fangataufa, the rate of growth was about 0.48 cm per year for the submarine stage and 0.10 after the volcano rose above sea level. By comparison the rates of growth of reefs are between 0.0012 and 0.0098 cm per year, much slower than the rate at which volcanoes can grow but fast enough to compensate for subsidence.

9.5 ▲ Hydrovolcanism on Land

Phreatomagmatic eruptions are especially common near the shores of islands. The picturesque promontory of Diamond Head near Honolulu was formed when basaltic magma rose through water-saturated sand and coral near the shore of Oahu. The ring of explosive ejecta owes its

(a)

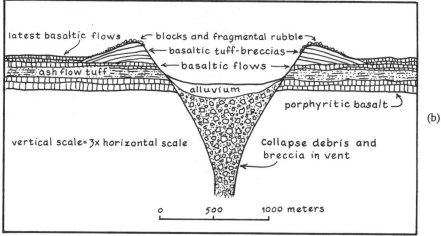

latest basaltic flows   blocks and fragmental rubble

basaltic tuff-breccias

basaltic flows

ash flow tuff

alluvium

porphyritic basalt

vertical scale=3x horizontal scale

Collapse debris and breccia in vent

0   500   1000 meters

(b)

Fig. 9.8 (a) The crater known as "Hole-in-the-Ground" was formed by eruptions through a shallow Pleistocene lake in central Oregon. It formerly contained a small lake. **(b)** A simplified cross section is based on drilling and on magnetic and gravity surveys by C. K. Kim. (**a,** Photograph by A. R. McBirney; **b,** Oregon Department of Geology and Mineral Industries.)

continued

(c)

Fig. 9.8 continued. (c) A typical maar in southwestern Germany. (c, Photograph by V. W. Lorenz.)

golden color to alteration (palagonitization) of the hyaloclastites when hot water and steam rose through the quenched basaltic glass.

Not all cones near the shores of islands are formed by new vents; some are the result of lava flows entering the sea. When fluid pahoehoe lava crossing the shoreline is fed by a well-defined channel or lava tube, the "rootless" crater that results from sustained phreatomagmatic explosions has many of the features of a normal adventive cone. *Litoral cones* of this kind are common on many oceanic islands. Craters can also be formed by secondary explosions when lava passes over wet ground or even by heavy rain. After the eruption of Pinatubo in 1991–1992, secondary phreatic explosions were set off by rain falling on the still-hot pyroclastic flows. The column of ash and steam reached a height of 18 km on the 21st of September in 1992, and 10 km three years later on the 11th of July in 1995.

On the continents, most hydrovolcanic eruptions occur in alluvial basins or beneath glaciers. A typical example of the former is the *tuff ring* of Hole-in-the-Ground, Oregon, which formed in a shallow lake that covered the area during Pleistocene time (Fig. 9.8). Other craters in the same region were filled by lava that rose as a coherent mass after the level of the rim rose above water level. Under such conditions, a wide spectrum of eruptions—including the directed blasts, pyroclastic flows, and surges described in Chapter 8—is possible depending on the amount of meteoric water the magma encounters.

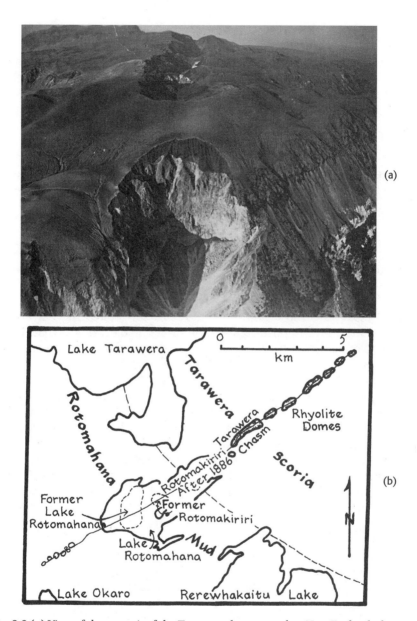

Fig. 9.9 (a) View of the summit of the Tarawera dome complex, New Zealand, showing the line of craters formed by the fissure eruption in 1886. A thick mantle of dark basaltic scoria covers the white rhyolite making up the main mass of the domes. **(b)** Plan view of the Tarawera fissure. Eruptions at lower elevations in the vicinity of Lake Rotomahana produced little fresh material but blanketed the surrounding region with mud and lithic debris. At higher elevations on the well-drained slopes of Tarawera, only basaltic scoria was erupted. (**a**, Photograph by S. J. Blackmore.)

Nowhere is this seen more dramatically than in the 1886 eruption of the New Zealand volcano Tarawera (Fig. 9.9). The eruption began when a fissure opened along a line of older domes crossing the summit. Within two hours, the fissure extended down the southern flank of the volcano and under an area of hot springs around Lake Rotomahana. The upper part of the fissure discharged fresh basaltic scoria, but where it crossed the hydrothermal basin, the eruptions were radically different. Instead of fresh magma, about a cubic kilometer of mud and lithic debris was erupted, and a swarm of explosion craters pitted an area 2 by 5 km across. Opening of the fissure had allowed the superheated water of the thermal area to flash to steam with devastating force. This was augmented by water from the lake that poured into hot levels of the fissure. The absence of fragments of basaltic glass in the erupted mud suggests that the descending water never came into direct contact with fresh magma.

The form of explosion craters depends in large measure on the state of the steam driving the eruptions. Most tuff rings that result from "dry" eruptions in which the water vapor remains in a superheated state (i.e., above 100°C at atmospheric pressure) have finely stratified, weakly indurated deposits and gentle slopes, usually of less than 12 degrees. This is in contrast to tuff cones, which result from "wet" eruptions in which part of the water is in the condensed, liquid state. The deposits have characteristics intermediate between those of Surtseyian and Strombolian eruptions; their particle sizes tend to be coarser and more compact, and their shapes have angles as sharp as 30 degrees.

The character of shallow eruptions can vary dramatically according to the level at which water gains access to the magma. Depending on the permeability of the wall rocks, water may enter the vent at different levels, so the focus of explosions may be shallow or deep. This, in turn, has an effect on the form of the crater and rim. Vents can range up to more than 1,500 m in diameter and 200 m in depth. As the vent is enlarged and taps deeper levels, the eruption may change from purely phreatic to phreatomagmatic. Most pipes formed in this way merge downward into dikes a few hundred meters below the surface, but a few remain fragmental to depths of 2,000 m or more. These deep *diatremes* rarely exceed 300 or 400 m in diameter, but they can have vertical dimensions measured in kilometers (Fig. 9.10, pp. 160 and 161). Although they are initiated by downward-propagated hydromagmatic eruptions, magmatic gases, particularly CO_2, also may play a role. The coarse breccia that fills them is composed of fragments from all levels of the walls and may also include exotic rocks and even diamonds from deep levels of the crust and mantle. Similar pipes can be seen in the Navajo region of northeastern Arizona and in the Eifel district of western Germany.

The prehistoric eruptions at Laacher See in the East Eifel district of Germany provide one of the best documented examples of tuff cone formation. A series of Plinian and phreatomagmatic explosions left a deep crater with a volume of 5 km³, part of it occupied by a lake. (Water-filled craters of this kind are referred to as *maars*.) The total erupted volume (the equivalent of 5.3 km³ of phonolitic tephra and 0.7 km³ of lithic clasts) corresponds roughly to that of the crater. Continual emissions of CO_2 indicate that a certain level of activity still persists. Another maar, Ulmener, was formed in the western part of the Eifel district only 10,000 years ago. Although there have been no historic eruptions, the activity in this region may not be over.

9.6 ▲ Eruptions through Lakes

Crater lakes have a wide variety of thermal and chemical regimes governed largely by their volcanic settings. Some, such as Lake Kawah-Idjen on the island of Java, contain large amounts of acid water; others, such as Lake Nejapa, Nicaragua, are exceptionally alkaline. Kelimutu, on the island of Flores, has two adjacent lakes: one jade green and the other brick red. The underwater fumaroles are oxidizing in the case of the former, but not the latter.

The temperatures of lakes may remain close to the boiling point for prolonged periods without erupting. The "Boiling Lake" on the island of Dominica and a lake of mud on Saint Lucia have temperatures of almost 100°C. Other lakes, however, become hot enough to erupt. The lake in the Costa Rican volcano Poas has gone through periods earlier in this century when the water rose like a geyser in periodic eruptions (Fig. 9.11). When magma rises into a deep, water-filled crater, it may simply produce mild steam eruptions, but as water is expelled and more of the lake is vaporized, the eruptions become increasingly violent. Eventually, after the water has been essentially eliminated, the eruption may take on a Strombolian character, as it did, for example, during the 1979 eruptions in the summit crater of the Soufrière of St. Vincent.

9.7 ▲ Subglacial Volcanism

Eruptions under ice resemble those under the deep sea, but they have a number of unique effects. Large amounts of meltwater can be formed beneath an ice cap, and when this water becomes thick enough to lift the ice, a sudden flood is released. The magnitude of the floods of water, ice, mud, and rock (or *jökulhlaups* as they are called in Iceland) is difficult to imagine. The eruption of Katla volcano in 1918, for example, caused a flood of about 100,000 m³ per second racing down the slopes and across the coastal plane. This is about half the rate of discharge of the Amazon, the largest river in the world. The floods were 70 m deep in places, and the debris they carried blanketed an area of 125 km². The 1934 eruption of Grimsvötn volcano, beneath the ice-cap of Vatnajökull, lasted only three days, but it produced 8.3 $\times$ 10⁹ m³ of lahars. The deposits were up to 8 km in width and had an average thickness of 2 m.

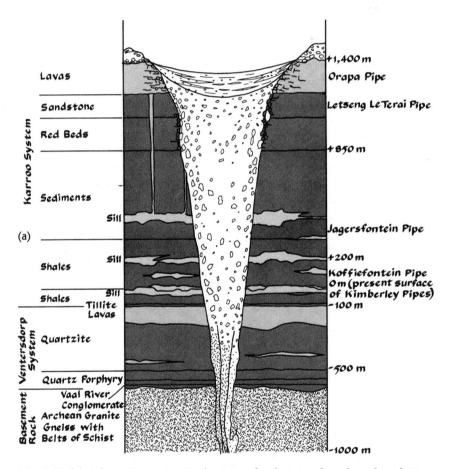

Fig. 9.10 (a) Schematic composite diagram of a diatreme based on the relations found in South Africa diamond mines. The pipes are of Cretaceous age and have been eroded and weather to different depths. The narrow feeder consists of multiple, unvesiculated dike-like bodies of differing compositions. (**a**, After Cox, A. G. 1978. *Sci Am* 238(4):120–32.)

The Vatnajökull is a plateau glacier about 140 km long, 90 km wide, and 2 km thick. Six volcanoes and high-temperature fumaroles are known to be buried under the ice. Because the heat from these vents produces large amounts of meltwater, the glacier has earned the name "glacier of water." Toward the end of September 1996, activity was detected on a well-known fault about 20 km long between the volcano Bardarbunga on the north and the lakes of Grimsvötn to the south. At the beginning of October, a fissure 3 km long and 300 m wide crossed the glacier. It was filled with a mixture of meltwater and volcanic scoria. Some days later a crater, a kilometer in diameter, was formed on the north side of the fissure. It released a continual column of steam several kilometers in height. Every 2 or 3 minutes, eruption clouds containing shards of black glass were ejected to heights of several

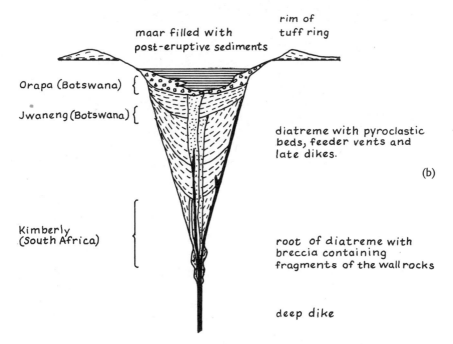

Fig. 9.10 continued (b) Schematic diagram of a diatreme-maar system. The deep part has been reconstructed from observations of the maar Orapa and the diatreme of Jwaneng in Botswana and the diatremes of the Kimberley region of South Africa. The horizontal and vertical scales are the same, but the widths of the pipe and dikes have been exaggerated. (**b,** after Lorenz, V. W. 1986. *Bull Volc* 48:265–74.)

Fig. 9.11 Eruptions through the crater lake of Poas volcano in Costa Rica eventually expelled all the water and took on a more Strombolian character. The lake, which returns during intervals of inactivity, contains large amounts of native sulfur. (Photograph by A. R. McBirney.)

hundred meters (Fig. 9.12). When the eruptive crisis ended in mid-October, 0.3 km³ of magma had been discharged as ash. Less than a month later, on the 5th of November, the glacier released a jökulhlaup—a gigantic flow of water, ice, and mud. Fortunately, the floods did not pass through densely inhabited areas.

One can easily recognize ancient subglacial volcanoes long after they have been uncovered. They are characterized by a flat top and steep slopes where their flanks were formerly encased in ice. The famous Herdubreid (Fig. 9.13) is a good example. It began its growth by erupting pillow lavas and hyaloclastites under a lake of meltwater, but when its summit finally emerged above the ice and water, it discharged

Fig. 9.12 Eruption under the Vatnajökull on October 9, 1996. A mushroom-shaped plume of black ash was ejected at high velocity along with a white cloud of steam. This is typical of phreatomagmatic eruptions caused by contact of water and magma. Note the concentric fractures of the glacier around the crater. (Photograph by J. M. Bardintzeff.).

Fig. 9.13 The Herdubreid of Iceland, a Pleistocene subglacial volcano from which the cover of ice has been removed. (Photograph by J. M. Bardintzeff.)

subaerial lavas that protected the summit from erosion after the ice had disappeared (Fig. 9.14). These flat-topped volcanoes are often called "table mountains," but those of the northwestern United States and Canada are usually referred to by their Indian name, *tuyas*.

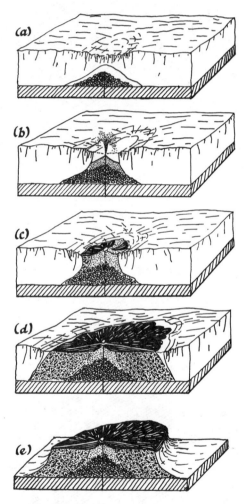

Fig. 9.14 Stages in the formation of a table mountain. **(a)** An eruption of basaltic pillow lava begins beneath a glacier and **(b)** melts the overlying ice. **(c)** The vent approaches the surface of the meltwater and begins to produce large amounts of hyaloclastites. When the vent emerges above the meltwater, it becomes less explosive **(d).** Lava forms a resistant cap over the underlying fragmental rocks, so that a flat-topped mountain is left when the ice disappears **(e).**

Suggested Reading

Crandell, D. R. 1971. Post-glacial lahars from Mt. Rainier volcano, Washington. *U S Geol Surv Prof Paper 677*, 75 p.
A description of the huge lahars that have descended into the region of what is now the city of Tacoma.

Self, S., and R. S. J. Sparks. 1978. Characteristics of widespread pyroclastic deposits formed by the interaction of silicis magma and water. *Bull Volc* 41:196–212.
Descriptions of the fragmental material produced by hydromagmatic volcanism.

Sheridan, M. F., and K. H. Wohletz. 1983. Hydrovolcanism: Basic considerations and review. *J Volc Geoth Res* 17:1–29.
A thorough review of the physics of magma–water interaction.

Simkin, T., and R. S. Fiske. 1983. *Krakatau 1883: The volcanic eruption and its effects.* Washington, DC: Smithsonian Institution, 464 p.
An account of the role of water in one of the most famous eruptions of historic times.

Swanson, D. A., S. D. Malone, and B. A. Samora. 1992. Mount Rainier: A decade volcano. *EOS* 73:171–8.
An excellent description of what is said to be the most dangerous volcano in the Cascades.

CHAPTER 10

Cones, Craters, and Calderas

▼

The thousands of volcanoes that contribute so dramatically to the earth's landscape have a seemingly infinite variety of morphological forms, ranging from towering cones, like Etna or Mt. Hood to broad, shallow depressions like Yellowstone. These differences reflect the great diversity of volcanic activity and are readily explained in terms of the eruptive processes described in the foregoing chapters.

10.1 ▲ Morphologies and Rates of Growth of Cones

The great majority of volcanoes follow broadly similar patterns of evolution. Although many factors contribute to their sizes and shapes, the most important are the viscosity of their lavas and their proportions of pyroclastic ejecta. Bombs, coarse scoria, and very viscous lavas tend to accumulate close to the vent and produce steep-sided structures, whereas fluid lavas move to more gentle slopes near the base, and fine ash is distributed as thin beds over wide areas.

Most growing cones pass through three stages of evolution. They form initially as scoria or spatter cones with slopes that are convex upward and lava flows that issue from vents near their base. As they develop further, pyroclastic ejecta are more widely distributed, lavas are erupted from the summit vent, and the slopes become concave upward. Finally, in the most mature stages, satellite cones and domes may form on the flanks, eruptions become less frequent but more explosive, and a caldera may form at the summit. In subduction-related calc-alkaline cones, this last stage is often marked by a divergence in the composition of the magma from monotonously uniform andesite to more or less contemporaneous basalts and rhyolite or dacite. As eruptions become less frequent, the constructional processes of volcanism fail to keep pace with the destructive processes of erosion, and the morphology is progressively modified until little of the structure's original form remains.

Many cones, especially the more familiar basaltic and andesitic varieties, are formed by eruptions that begin from fissures and later become centered in a single central vent. The birth and development of the volcano Paricutin 300 km west of Mexico City is a prime example. On the 20th of February 1943, a Mexican farmer, Dionisio Pulido, observed that a small fissure that had existed for many years in his corn field had become a "smoking crevasse" and soon was ejecting scoria to form a rapidly growing cone. It reached a height of 10 m after 12 hours, 30 m after the first day, 106 m at the end of a week, 148 m after a month, and 336 m by the end of the year. Starting in July 1943, lavas emerged from the base of the cone, and the following year, they buried all of the neighboring village of San Juan Parangaricutiro except a church steeple that still rises above the blocky aa flows. When activity came to an end on the 4th of March 1952, the summit crater of the cone was 424 m high.

Throughout their period of rapid growth, the shapes of scoria cones remain almost constant. Slopes reach a maximum *angle of repose* of about 30 degrees, which is the steepest angle at which loose scoria and other granular material is stable. Steeper slopes may develop if bombs are still hot when they land and are plastic enough to become welded. Some of these spatter cones may be as high as 30 m and 100 m in diameter.

Strombolian cones like Paricutin are common in many parts of the western United States. SP Crater in northern Arizona and Lava Butte, Oregon, are good examples (Fig. 10.1). They tend to occur in clusters of dozens or more scattered over areas of a few tens to hundreds of square kilometers. Many seem to be *monogenetic* in the sense that, like Paricutin, they are products of a single eruption of scoria and lava. Only a few go on to become large cones.

Certain volcanoes, such as Stromboli (Italy), Sangai (Ecuador), and Yasour (on Vanuatu, New Hebrides), have been more or less continuously active for centuries with mild explosions at intervals ranging from a few minutes to hours. Stromboli has been in this state of activity for thousands of years without major changes of its geometric form or type of magma. The rate at which lava and scoria are discharged requires a steady supply of about 1 kg of magma per second, but the amount of gas emitted year after year requires much more, possibly as much as hundreds of kilograms per second. The amount of

Fig. 10.1 (a) SP Crater in the San Francisco Volcanic field of Arizona is typical of many monogenetic cones in the western United States. Note the trace of a fault crossing the upper-left side of the photo. **(b)** The cone of Lava Butte in central Oregon. Note that the scoria and lava were erupted simultaneously from separate parts of a fissure running under the cone. No lava was erupted from the summit crater. (**a,** Courtesy of Wendell Duffield, U.S. Geological Survey.)

magma discharged must be only a small fraction of the total input into a shallow reservoir, but it is difficult to account for the large difference between the erupted magma and the amount needed to supply the huge quantities of gas.

10.2 ▲ Mature Composite Cones

Cones that continue to have frequent, recurrent eruptions can rapidly take on the form of mature composite volcanoes. The volcano Izalco in El Salvador is one of the few volcanoes that first erupted in historical time and has been observed to build a large cone. After its formation in 1536, it has erupted lava and scoria every few decades, and after two centuries it had a total volume of about 1 km³. During these later eruptions, lava was discharged from the summit crater rather than from the base. The volcano has now reached a height of about 500 m, and if it continues at this rate, it will resemble the great cones in adjacent parts of the Central American chain.

Virtually all major continental volcanoes are "composite" in the sense that they are built partly of lava, partly of fragmental ejecta, and partly of shallow intrusions. Although these *stratocones* form some of the earth's most majestic landforms (Fig. 10.2), they are small compared with many oceanic volcanoes. The summit elevations of many imposing cones, such as those of the Andes, exceed 5,000 m above sea level, but this is because they are constructed on a high mountain range; few rise more that half that height above their base.

The shapes of composite cones differ widely; no two are exactly alike. When the summit reaches a certain height, eruptions of lava tend to break out from radial fissures on the flanks, even though explosive eruptions may continue from the summit. This change to flank eruptions comes about when the height of the magma column becomes so great that the pressure on its walls opens fissures and allows the magma to escape laterally at a lower elevation. Mt. Etna, one of the world's largest land volcanoes, has 250 of these *adventive cones*. The largest, Monti Rossi, was built during the eruption of 1669 and measures 250 m in height.

As activity becomes less frequent, eruptions tend to be more violent. In many cases, they result in a caldera or amphitheater that may contain one or more scoria cones and domes, such as the one that has grown in the summit depression of Mt. St. Helens. Meanwhile, erosion removes debris from the flanks and redeposits it around the base, thus contributing to the steady steeping of the upper slopes. The structure of a typical cone can be divided into three zones, each with a distinctive form and assemblage of rocks (Fig. 10.3). The *central zone*, within

Fig. 10.2 (a) The large volcanoes of Kamchatka are excellent examples of the symmetrical composite cones that characterize the Circum-Pacific andesitic chains. Note the viscous lava flows and small radially aligned satellite vents. **(b)** The cone of Toliman in Guatemala shows the effect of viscous lava flows that have accumulated around the lower slopes of a steep central cone composed mainly of pyroclastic material. (**a,** Courtesy of G. S. Gorshkov; **b,** photograph by U.S. Air Force.)

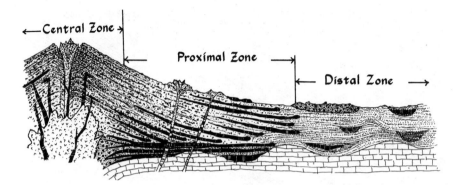

Fig. 10.3 A schematic section through a typical andesitic volcano of the Cascade Range shows the spatial relationships of various types of volcanic rocks at increasing distances from the central vent. The plutonic rocks shown intruding the central core are a feature seen in many deeply eroded volcanoes of this kind.

about 0.5 to 2 km of the vent, is distinguished by some combination of the following:

- Coarse Strombolian ejecta
- Vent breccia
- Radial dikes
- Thick, steeply banded siliceous lavas
- Ponded crater fillings

The *proximal zone,* up to 5 to 15 km from the central vent, has many of the following types of deposits:

- Broad, thick lava flows
- Lahars with large angular blocks
- Well-sorted tephra
- Strongly welded ignimbrites
- Clastic debris reworked by water

The *distal zone,* more than 5 to 15 km from the vent, has deposits with greater lateral continuity than those of inner zones. They normally include the following:

- Finely layered ash
- Lahars with few large blocks
- Ignimbrites with moderate to weak welding
- Interlayered shallow-water sediments
- Lava flows restricted to outlying vents and intracanyon flows

These facies relations can be very helpful when one is trying to use incomplete exposures to reconstruct the form of an ancient, deeply eroded volcano.

10.3 ▲ Shield Volcanoes

Volcanoes that erupt fluid basaltic lavas with relatively small proportions of pyroclastic ejecta take on a much more gentle profile than andesitic cones, such as those of the Cascade Range and Japan. They were first referred to as shield volcanoes in Iceland where their shape resembled that of a Viking warrior's shield. Although a few volcanoes with this distinctive form are found on the continents, they are much more common in the oceans. The great volcanoes of Hawaii are probably the most familiar examples.

Like the cones described in the preceding section, the evolution of Hawaiian volcanoes can be traced through three main stages. The earliest submarine phase is represented by a small cone, Loihi, growing near the southern base of the island of Hawaii. Frequent eruptions of pillow lavas from a complex of several vents have resulted in irregular slopes around a complex of vents. With time, its form will become more regular as eruptions become centralized in a single vent. In order to reach the surface and emerge as an island, a submarine volcano like Loihi must produce huge volumes of magma in the form of lavas, dikes, and sills. Only a small proportion of the cones formed on the seafloor reach this stage, and even fewer survive the passage from subaqueous to subaerial conditions. As the eruptive vent approaches shallow depths, the enhanced explosive effect of heated seawater produces large amounts of fragmental debris, and unless this weak material is armored by more resistant lava flows, it is easily eroded by the action of waves (see Chapter 9).

Explosive eruptions continue until the shield emerges well above sea level, but they become much less important than lava. Thin flows traveling long distances down the flanks produce slopes as gentle as 45 to 65 m per kilometer. Traveling from the coast to the summit of Kilauea thousands of feet above sea level, one fails to appreciate the enormous size of these shields because the slopes are so gentle.

Mauna Kea, the highest of the Hawaiian volcanoes, has grown to an elevation of 4,206 m in less than a million years, and if measured from its base 4,000 m below sea level, it is about as high as Mt. Everest—the highest mountain on earth. The volume of its neighbor, Mauna Loa, is almost 300 times that of Fujiyama—one of the largest volcanoes on land (Fig. 10.4). Large as it is, however, Mauna Loa is small compared with the giant shield volcanoes of other planets (see Chapter 12).

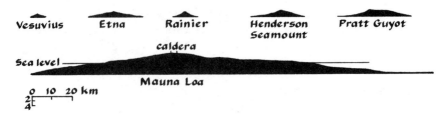

Fig. 10.4 Profiles drawn at the same scale through various large continental and oceanic volcanoes.

It was long thought that calderas developed only late in the subaerial shield-building stage, but the recent recognition of a circular depression on the summit of Loihi suggests that they may form even in the earliest submarine stages. Radial rifts are characteristic of the Hawaiian shields but not of all shield volcanoes; those of the Galapagos Islands have concentric fissures, and some of the giant shields of East Africa have none at all.

In the case of the Hawaiian shields, the final stages of declining activity are characterized by scattered eruptions from scoria cones and trachytic domes near the summit and, less commonly, well down the flanks. With time the caldera may be completely filled. The last magmas erupted before the volcanoes become extinct are highly alkaline basalts containing silica-deficient minerals, such as nepheline and melilite.

10.4 ▲ Craters

Craters take on a variety of sizes and forms, depending on the physical properties of the rocks and the manner in which they are erupted. Most craters at the summits of pyroclastic cones are not depressions excavated by explosive ejections but simply rims of ejecta surrounding a vent. In that sense, craters are actually constructional landforms. Blocks and large bombs thrown out by relative weak explosions accumulate as a steep rampart close to the vent, whereas finer ejecta propelled by strong explosions produce a broader crater with more gentle slopes. Wind direction has a marked effect on symmetry. If it carries the ejecta predominantly in one direction, the rim is larger on the downwind side and in extreme cases can have the shape of a crescent or horseshoe.

The broad, shallow depressions or *tuff rings* left by phreatomagmatic eruptions (see Chapter 9) are true craters in the sense that they are produced by explosive excavation of the subsurface (Fig. 10.5);

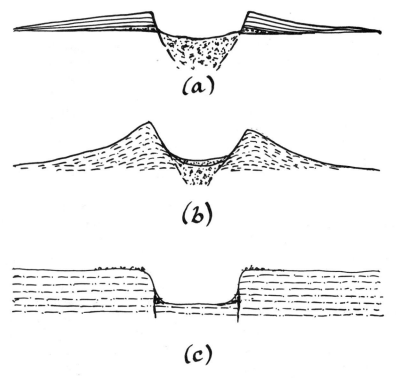

Fig. 10.5 (a) Tuff rings have broad rims surrounding a crater with a floor that is at or below the surrounding ground level. If the floor is below the ground water level and the crater is occupied by a lake, it is called a maar. **(b)** Tuff cones tend to have steeper sides and a crater floor above the surrounding ground level. **(c)** Pit craters are formed by collapse of a cylindrical blocks along ring faults. There may be little or no explosive ejecta on their rim.

Hole-in-the-Ground in central Oregon is a typical example (Fig. 9.8). Tuff rings containing a lake are referred to as *maars*.

Vents that discharge lava with little accompanying pyroclastic material rarely have conspicuous craters, but cylindrical depressions can form when magma withdraws from the vent and the overlying crust drops into an evacuated space. These *pit craters* are easily distinguished from explosion craters by their steep walls and the almost total absence of a pyroclastic rim. They are especially common on rift zones, such as those of the Chain of Craters on the East Rift Zone of Kilauea. When formed on rift zones, the collapse may not necessarily be caused by a withdrawal of magma. Recent work in Hawaii suggests that magma flowing beneath the surface stops blocks from the roof and carries them downstream. Rather than collapsing into a void, the thin crust drops into the underlying current of magma.

10.5 ▲ Calderas

Large depressions resulting primarily from collapse have been given the name *caldera,* taken from the Portuguese word *caldeira* meaning "cauldron." Calderas resemble pit craters in that both are formed by collapse, but calderas are larger, generally more than 1 or 2 km in diameter. Although craters and small calderas are funnel shaped, larger depressions have steeper boundary faults around a relatively flat, circular, or elliptical floor. Multiple stages of piecemeal collapse produce scalloped walls, but in most very large calderas the shape and alignment of the boundary faults reflect structural features of the underlying crust.

Calderas have been divided into a variety of types, but the most important distinction is one based on the composition of the magma, which may be either mafic or felsic. Although both types are formed by collapse following evacuation of an underlying magma chamber, they differ widely in form and eruptive style.

The large summit depressions of Hawaiian volcanoes, such as Kilauea and Mauna Loa, are classic examples of mafic calderas (Fig. 10.6a). They were not formed by a single event but by piecemeal subsidence over many centuries. Depression of the floor normally follows copious discharges of basaltic lava from radial rifts extending far down the flanks of the shield. Drainage through radial rifts is not essential to their formation, however. The Masaya caldera of Nicaragua, for example, does not occupy the summit of a large shield but rather a volcanic center of low relief with few if any eruptions outside the boundaries of the caldera. In such cases, collapse may result from lateral injections of sills or from recession of magma from a broad, shallow reservoir.

In contrast to the calderas of mafic volcanoes, those of the felsic type are formed by voluminous eruptions of siliceous pyroclastic material. The 1883 eruption of Krakatao is by far the best documented example of a caldera-forming event of this kind. The eruption (described in Chapter 7) resulted in collapse of the central part of the volcano into the waters of the Sunda Straits and generated huge tsunamis along the adjacent shores of Sumatra and Java. The volume of ejecta was estimated at 9 to 10 km^3, of which 95% was new rhyodacitic magma and the rest older rock. This is substantially more than the volume of the volcano that disappeared (6.74 km^3).

The eruption that produced the caldera of Crater Lake in the summit of Mt. Mazama, Oregon (Fig. 10.6b), was similar to that of Krakatao but considerably larger. It occurred about 6,850 years ago following a rapid discharge of about 55 km^3 of rhyolitic pumice (Fig. 10.7). As in the case of Krakatao, the initial Plinian eruptions changed to pyroclastic flows, and toward the end, the composition of the magma became more mafic and crystal rich. The volume of the caldera, which measures 10 km across and 1,200 m deep, together with that of the summit that disappeared, is about 60 km³ or slightly more than the estimated volume of erupted magma.

(a)

(b)

Fig. 10.6 (a) Aerial view of Mokuaweoweo on Mauna Loa, Hawaii. **(b)** The caldera of Crater Lake at the summit of Mt. Mazama, Oregon, was formed about 6,850 years ago following a major eruption of pumice that evacuated the underlying reservoir of siliceous magma. (**a,** Photograph by U.S. Air Force; **b,** from National Parks Service.)

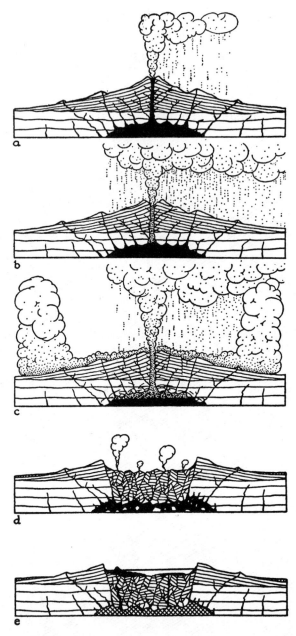

Fig. 10.7 The sequence of events responsible for formation of Crater Lake was typical of most caldera-forming eruptions. An initial eruption **(a)** opened a vent from which increasing volumes of magma were discharged in a large Plinian column **(b).** Pyroclastic flows swept down the flank **(c),** and lowering of the level of magma in the reservoir undermined the summit of the volcano. The entire structure collapsed, forming the summit caldera **(d).** Later eruptions formed cones and domes on the caldera floor and accumulating meteoric water created a lake.

Most calderas in volcanic arcs have diameters of less than 15 km (Fig. 10.8), but those in the interiors of continents tend to be much larger. The Valles Caldera of New Mexico, for example, measures 20 by 25 km, and that of Yellowstone, about 45 by 75 km. Because of their great size, these calderas may be difficult to recognize on the ground, but they stand out clearly on maps and satellite images. The volumes of magma they have discharged are equally large. Yellowstone, for example, produced about 3,800 km³ of ignimbrites in three episodic caldera-forming eruptions 2.0, 1.2 and 0.6 million years ago (Fig. 10.8). The first produced 2,500 km³, the second about 1,000, and the third about 280. These huge volumes of rhyolitic magma are thought to come from melting of the continental crust by a plume of basaltic magma over which the American Plate is migrating at a rate of 4.5 cm per year (see Chapter 11).

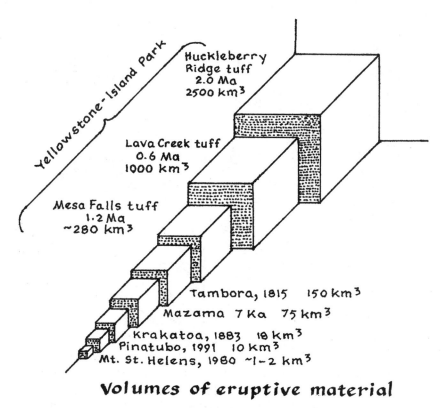

Volumes of eruptive material

Fig. 10.8 Volumes of pyroclastic material ejected by various volcanoes compared with that of the three main eruptive phases of Yellowstone. (Compiled by Smith, R. B., and L. W. Braile. 1994. *J Volc Geoth Res* 61:121–87.)

The Toba depression of Sumatra differs from other large felsic calderas in that it is clearly related to a regional structure known as the Semangka rift zone (Fig 10.9). The world's largest known caldera, it has an elliptical shape 100 by 30 km and is a composite of several coalescing basins parallel to the axial graben of the main volcanic chain. About 75,000 years ago, siliceous magma broke through the crust and in a series of closely spaced eruptions discharged about 2,800 km³ of magma, mainly in the form of ignimbrites that covered between 20,000 and 30,000 km². Ashfall deposits, amounting to another 800 km³, were laid down over much of Indonesia and Malaya. These huge outpourings took place over a surprisingly short interval of time, possibly 9 to 14

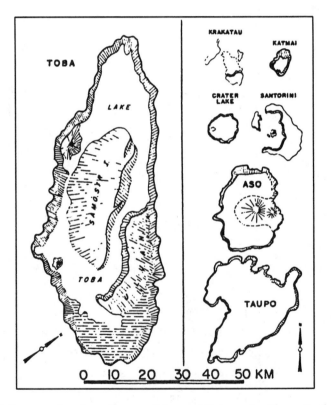

Fig. 10.9 Comparative sizes of some of the world's largest calderas. In the case of the Toba caldera of Sumatra (100 by 30 km), part of the 2-km deep depression is occupied by a lake. Doming of the floor created the island of Samosir in the middle. Its great size becomes evident when it is compared with other calderas shown at the same scale on the right. (After Ninkovitch, D., R. S. J. Sparks, and M. T. Ledbetter. 1978. *Bull Volc* 41–3:286–98.)

days. The ash has been found in cores of marine sediments in the Indian Ocean as far as 2,000 km from the volcano.

Although it is generally assumed that calderas are underlain by large intrusions, it is rarely possible to determine the structure beneath their floors. Deeply eroded calderas, such as those of the San Juan Mountains of Colorado, are associated with sizable granitic or dioritic plutons (Figs. 10.10 and 10.11). These seem to have fed vents

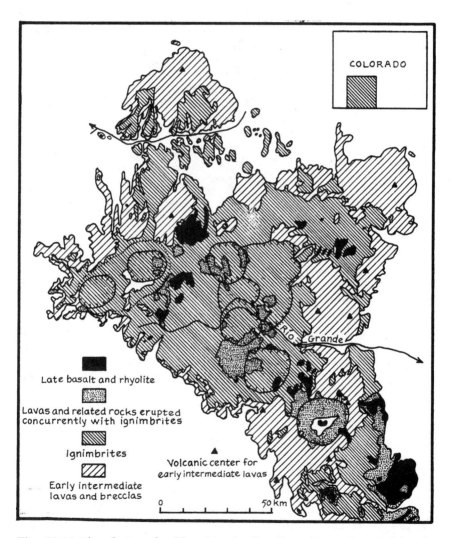

Fig. 10.10 The cluster of calderas in the San Juan Mountains of Colorado is related to granitic intrusions that rose beneath a complex of andesitic cones. (After Lipman, P. W. 1984. *J Geoph Res* 89:8801–41.)

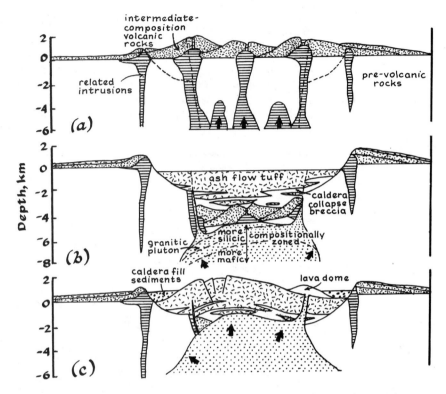

Fig. 10.11 The three stages of development of the calderas are illustrated in Fig. 10.10. The top diagram illustrates precollapse volcanism over small, high-level intrusions. New intrusions result in uplift *(heavy arrows)*, causing faulting on ring faults indicated by steep dotted lines. The middle diagram shows the form of the caldera just after collapse, following eruptions of great quantities of siliceous magmas. Slumping takes place on curved, inward dipping planes. In the lower diagram, postcaldera volcanism and doming are caused by rise of the main body of magma into the base of the subsided block. (After Lipman, P. W. 1984. *J Geoph Res* 89:8801–41.)

near the center and along the boundary faults of the caldera floor. Seismic studies of the caldera of Long Valley in southern California show that a body of magma with a diameter of about 10 km lies at a depth of 8 to 10 km below the surface (Fig. 10.12). This is only the part that is still largely molten; the size and shape of the entire intrusion are unknown. The chamber had a volume of about 3,000 km³ before its paroxysmal eruption about 700,000 years ago, but since that time, its volume has been reduced by about a third through a combination of crystallization and relatively small eruptions. Although the caldera

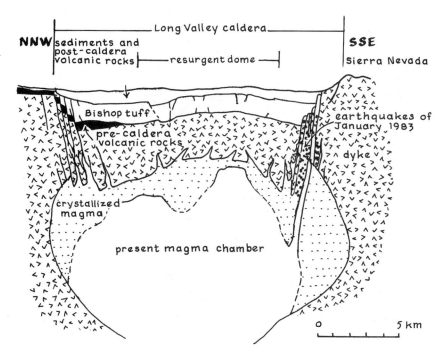

Fig. 10.12 Schematic cross section through the caldera of Long Valley and its underlying body of magma. The vertical and horizontal scales are the same. (From Hill, D. P., R. A. Bailey, and A. S. Ryall. 1985. *J Geophys Res* 90:B13, 11111–20.)

is still theoretically capable of discharging 30 km³ or so of magma, the volumes produced by eruptions in the last half million years have never exceeded about a cubic kilometer. A new influx of fresh magma could, of course, lead to a much larger eruption.

Most calderas have had repeated eruptions over tens or hundreds of thousands of years. During intervals of repose the reservoir of magma is replenished, the floor is elevated and small domes may be erupted near the center or along ring faults at the caldera boundary (Fig. 10.13). This effect is best documented in the Aegean volcano Santorini, which has had at least three major caldera-forming eruptions, the first about 180,000 years ago, the second 21,000 years ago, and the most recent around 1645 BC. In the intervening years, smaller eruptions have formed domes similar to those that are now active in the center of the caldera floor.

As new magma enters the system, the entire structure of these volcanoes begins to swell, but the uplift is not always followed by an immediate eruption. The floor of the Yellowstone caldera was uplifted

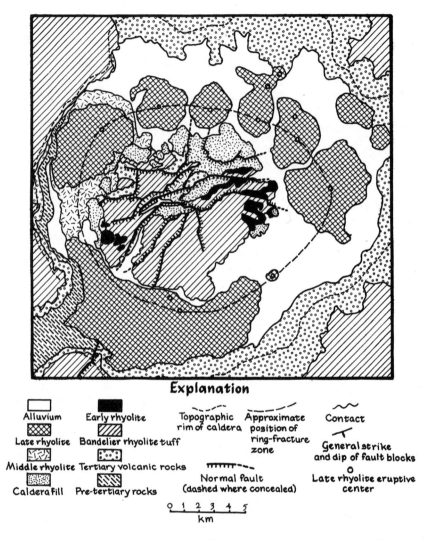

Explanation

☐ Alluvium	■ Early rhyolite	Topographic rim of caldera
⊠ Late rhyolite	▨ Bandelier rhyolite tuff	Approximate position of ring-fracture zone
Middle rhyolite	Tertiary volcanic rocks	Contact
Caldera fill	Pre-tertiary rocks	Normal fault (dashed where concealed)
		General strike and dip of fault blocks
		Late rhyolite eruptive center

0 1 2 3 4 5
km

Fig. 10.13 Domes within the Valles caldera of New Mexico are arranged in a circular pattern that seems to be controlled by the boundary faults of the floor.

by as much as a meter between 1923 and 1984 and then subsided as much as 12 cm from 1985 to 1991 without any discharge of magma at the surface. The Long Valley caldera began to swell at an alarming rate between 1975 and 1986 (Fig. 10.14), and numerous earthquakes were felt throughout the region, but the dreaded eruption has not yet materialized. As we shall see in a later chapter, events such as these present a serious problem for volcanologists responsible for assessing volcanic hazards.

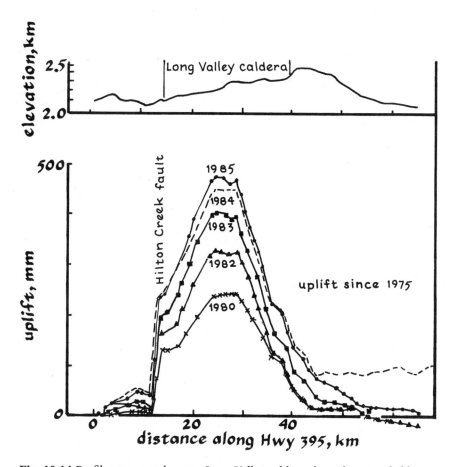

Fig. 10.14 Profiles measured across Long Valley caldera show the remarkable uplift of the floor between 1980 and 1985 relative to 1975. Upper diagram shows a topographic profile across the caldera. (After Bailey, R. A., and D. P. Hill. 1990. *Geosci Canada* 17:175–9; Savage, J. C., et al. 1987. *J Geoph Res* 92:2721–46. A complete review can be found in a special issue of the *Journal of Geophysical Research* [vol. 89, B10] published in September of 1984.)

Suggested Reading

Bacon, C. R. 1983. Eruptive history of Mount Mazama and Crater Lake Caldera, Cascade Range, U. S. A. *J Volc Geoth Res* 18:57–115.
   An up-to-date study of the caldera and the products of its eruptions.

Dzurisin, D., K. M. Yamashita, and J. W. Kleinman. 1994. Mechanisms of crustal uplift and subsidence at the Yellowstone caldera, Wyoming. *Bull Volc* 56:261–70.
   A study of the structural development of Yellowstone caldera.

Hill, D. P., R. A. Bailey, and A. S. Ryall. 1985. Active tectonic and magmatic processes beneath Long Valley caldera, Eastern California: An overview. *J Geoph Res* 90:B13, 11111–20.
A description and interpretation of ground movements during the recent unrest of a major caldera.

Luhr, J. F., and T. Simkin. 1993. *Paricutin, the volcano born in a Mexican cornfield.* Phoenix: Geoscience Press.
A compilation of studies of the most thoroughly documented example of the birth and development of a new volcano.

Williams, H. 1941. Calderas and their origins. *Univ Calif Publ Geol Sci* 25:239–346.
A classic work in which much of the basic concepts and terminology of caldera were first defined.

Williams, H. 1942. The geology of Crater Lake National Park, Oregon, with a reconnaissance of the Cascade Range southward to Mt. Shasta. *Carnegie Inst Wash Pub* 540: 162 p.
Another of Williams' classic works on calderas.

Volcanoes in a Geodynamic Context

▼

11.1 ▲ Global Distribution of Volcanism

The nature of magmatism in various tectonic settings differs widely in the amounts and compositions of erupted magmas, and in the proportions of extrusive to intrusive activity. Under most conditions, the greatest proportion of magma crystallizes at depth in the form of plutonic intrusions, but the ratio is far from constant. The ratios of intrusive to extrusive rocks are typically about 5 to 1 for oceanic spreading axes but more like 10 to 1 in continental settings. On the continents, the rates of production and eruption for basaltic magmas appear to be only marginally greater than for more siliceous compositions; for the oceans, all but a small fraction are basalt.

For the past century, the total rate of global magmatism, including both surface eruptions and subsurface intrusions, has been estimated at 26 to 34 km³ per year. Of this volume, about 75% has been produced at the oceanic ridges, 15 to 20% at subduction zones, 5% in the interior of oceanic plates, and less than 5% in the continental interiors.

11.2 ▲ Oceanic Ridges and Continental Rifting

The 60,000 km-long system of oceanic spreading axes is by far the major locus of magmatism on earth. We cannot normally observe this activity because it lies hidden beneath the seas, but much has been learned from submarine exploration and geophysical studies (Fig. 11.1). We can also examine the structures and types of rocks in places where a spreading axis emerges onto land, as it does in Iceland and

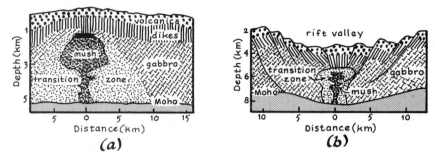

Fig. 11.1 (a) Cross-axis section through a fast-spreading ridge, such as the East Pacific Rise. The structure has been defined from the seismic properties as described in Chapter 2. Only minor faulting is seen along the axis, and magma tends to rise at widely spaced points and spread laterally as sills. **(b)** Slow-spreading ridges, such as the Mid-Atlantic Ridge, have medial grabens and only intermittent bodies of magma.

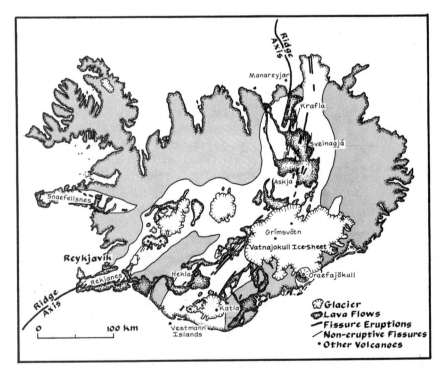

Fig. 11.2 The Mid-Atlantic Ridge emerges near the southwestern tip of Iceland and is offset toward the center of the island before disappearing again below the Arctic Ocean.

East Africa. In addition, the deeper levels of these systems can be seen in *ophiolite complexes*, where the oceanic crust has been thrust onto the continent and incised by erosion.

Where the Mid-Atlantic Ridge crosses Iceland, periodic fissure eruptions are linked to dilation of the crust at a rate of about 2 cm per year (Fig. 11.2). Much of this extension is taken up by dikes that fail to reach the surface. The crust above these dikes is cut by small grabens and gaping fissures. The most active region today is in the northern part of the island where the rift has been erupting basaltic lavas every decade or so, but much larger eruptions take place at intervals of centuries. The tenth century eruption of Eldgja produced about 15 km³ of lava from a fissure nearly 70 km in length, and comparable amounts poured from the parallel fissure of Laki in 1783.

The best known example of rifting in a continental setting is the East African Rift extending from the Red Sea and Afar Depression southward through Ethiopia, Kenya, and Tanzania (Fig. 11.3). Rifting was preceded by a broad upwarping of the surface that was soon followed by block faulting and fissure eruptions of alkaline basalts and siliceous ignimbrites. In Pliocene time, basaltic shield volcanoes became more common. Explosive eruptions of siliceous magma were confined mainly to the axis of the rift, while basalts poured out on the adjacent plateau.

As explained in Chapter 9, the characteristic products of volcanism on the seafloor are pillow lavas with various amounts of hyaloclastites. At the ridges, these lavas are fed by a system of dikes that increase in number downward until they grade into masses of gabbro with small, irregular bodies of slightly more differentiated rocks. The underlying mantle rocks are mainly harzburgites that seem to be the refractory residue left after extraction of a basaltic melt (see Chapter 1).

The magmas are a common type of olivine-bearing tholeiite usually referred to as MORB, for mid-ocean-ridge basalt. They have about 50% SiO_2 and are relatively poor in alkalies, especially potassium. Differentiated rocks are very subordinate. In contrast to the magmas of oceanic ridges, those of continental rifts are much more varied. In the eastern or Gregory Rift of East Africa, for example, the rocks tend to be sodic, whereas volcanoes along the western branch have produced some of the most potassic lavas known. Of the total of 220,000 km³ of Cenozoic volcanic rocks in the region of the Kenya Rift, about half are highly differentiated rocks, such as phonolites, trachytes, and rhyolites.

There is no way of knowing whether the East African Rift will continue to open and eventually produce an open sea, as the Mid-Atlantic Ridge did when North America split away from Europe. Many rifts begin to open and then, for reasons we do not understand, become inactive. The Rio Grande Rift of New Mexico is an example of a "failed rift" of this kind;

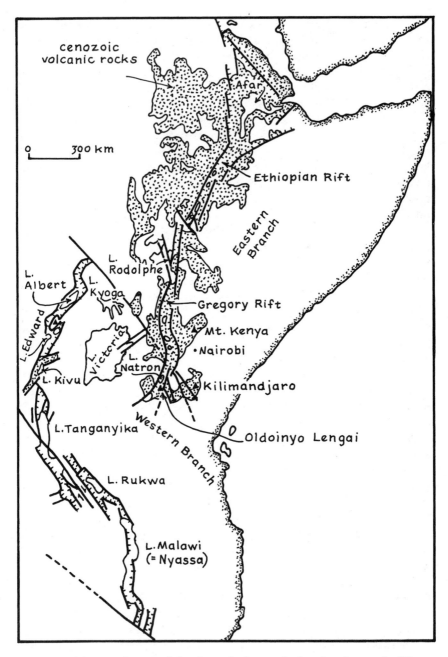

Fig. 11.3 The East African Rift has been the focus of volcanism for much of Cenozoic time. In the southern part of the system, volcanism is associated mainly with large central-vent volcanoes like Mt. Kilimandjaro, but fissure eruptions are more important in the northern section, especially in the Afar Depression. (From Chorowicz, J. 1983. *Bull Elf Aquit* 7: 155–62.)

the Rhine Graben is another. Both of these rifts have been the locus of recent volcanism, but the scale of activity seems to be declining.

Because the surface area of the earth is essentially constant, additions of crust at oceanic ridges must be compensated by consumption of an equal volume elsewhere. The balance is achieved near the ocean margins where the oceanic lithosphere is either *subducted* back into the mantle or, more rarely, *obducted* on to the continental margin.

11.3 ▲ Subduction-Related Volcanism

Volcanism at convergent plate boundaries is second in importance only to that of spreading axes. It can be divided into two general types, depending on whether the oceanic plate is subducted below an oceanic or continental one. Where both plates are oceanic, the surface manifestation of subduction is an island arc. A boundary where an oceanic plate descends below a continental plate is referred to as an *active continental margin.*

The boundaries between oceanic and continental plates are not necessarily convergent; the Atlantic coasts of North and South America, for example, are passive boundaries with no subduction and hence no volcanism. The Circum-Pacific chain of volcanoes, often referred to as the "Circle of Fire," includes both the continental margins of western North and South America and the island arcs of the northern and western Pacific rim. In parts of the western Pacific, such as Japan and New Zealand, the arcs have components of both continental margins and island arcs (Fig. 11.4).

The form of subduction-related boundaries is largely a function of the rate and angle of subduction and the type of lithosphere in the overriding plate. The more rapid the rate of subduction, the deeper the extent of earthquakes and the deeper the offshore trench, provided, of course, the trench is not flooded with sediments and volcanic debris from the adjacent islands or continent. Volcanism is focused along a *volcanic front* directly above a zone of melting about 150 to 200 km from the trench; lesser amounts appear in back-arc basins where local extension of the crust facilitates the rise of magma. The heights to which cones grow are about the same in island arcs and continental margins, but where the overriding plate is thin oceanic lithosphere, they rise from the ocean floor; volcanoes growing on thick continental crust start at higher elevations and can reach elevations of 5,000 to 7,000 m.

The most complete data on rates of subduction-related volcanism are those compiled for Central America and the Antilles, where it is possible to estimate the volumetric rate of production of magma per kilometer length along the volcanic chains. For the past 300 years, the

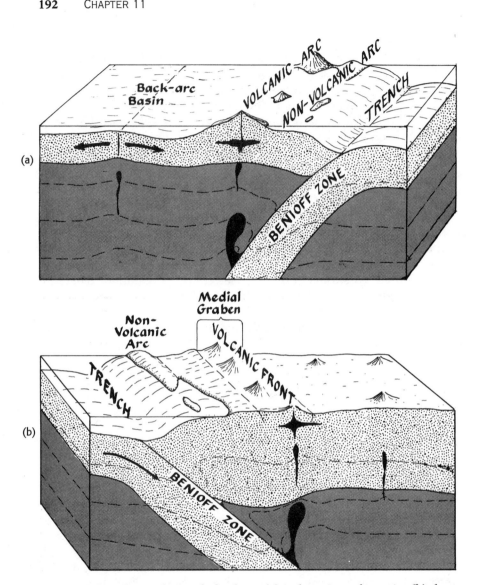

Fig. 11.4 The volcanic chains of island arcs **(a)** and continental margins **(b)** share many essential structural elements. A trench, which in some places may be filled with sediments, marks the surface expression of subduction. Immediately inward and parallel to the trench a non-volcanic arc or coast range consists of uplifted and often highly deformed older sedimentary and igneous rocks. A second, less conspicuous, uplift rises to the main volcanic front where large andesitic cones rise within or along the margins of a medial graben. Volcanoes may be distributed behind the volcanic front, but they are smaller and fewer in number. Most island arcs have a back-arc basin that seems to result from spreading behind the main volcanic arc.

average productivity has been about 62 km³ per kilometer per million years in Central America and 5 km³ in the Antilles (compared with about 300 km³ for mid-oceanic ridges). These rates are far from constant. Over the past 100,000 years, they have declined to 31 and 4 km³, respectively. Throughout this time, the ratio of the two values has varied between 8 and 12, suggesting that the greater activity of volcanoes along the continental margin of Central America is not a random feature. The rates of subduction are estimated at 9 cm per year for Central America and only 2 cm per year for the Antilles, giving a ratio of about 4. Thus, there appears to be a direct correlation between subduction and production of magma and, on a larger scale, between tectonics and volcanism. This relationship between rates of magmatism and subduction is particularly clear in the Aleutian Islands, where the amount of magma discharged diminishes from east to west along the length of the arc in the same direction as the rate of subduction declines.

Rates of volcanism change in response to any variation in the relative motion at plate boundaries, and in some instances, these variations seem to occur simultaneously over wide regions. Most of the large, presently active volcanoes of the Circum-Pacific first began to erupt at the same time about a million years ago, and looking back into the earlier history of these volcanic chains, one finds similar episodes in which volcanism appears to have increased or decreased simultaneously over large parts of the earth. The time scale of these changes is measured in hundreds of thousands to millions of years. Volcanism seems to appear between 1 and 3 million years after subduction begins and continues for 10 million years or so after it comes to an end.

Geophysical evidence indicates that partial melting probably occurs at depths of about 100 km, where fluids liberated from the descending slab of oceanic lithosphere rise into the overlying mantle and lower its melting temperature. As we noted in Chapter 1, the magmas produced by this process are relatively rich in silica and alumina and poor in iron and titania. Their compositions vary laterally across the arc, often with an increase of potassium in rocks farther inward from the trench. They tend to be more tholeiitic at the fore-edge, calc-alkaline in the main volcanic chain, and alkaline (shoshonitic) in the far interior. The increase of potassium has been correlated with the age of subduction and with the depth of the underlying part of the Benioff Zone, but it is also related to the thickness and composition of the crust through which the magma rises. Most subduction-related magmas assimilate components of the crust through which they rise; in places, they seem to be produced by melting of the walls and roofs of deep magma chambers. The sillimanite-bearing ignimbrites of Peru and Chile have been attributed to melting under such conditions in the deep roots of the Andes.

11.4 ▲ Collision-Related Volcanism

Because the continental lithospere has a much lower density than the mantle, it resists being subducted. When two continental plates collide, a limited amount of convergence can be taken up by folding and faulting, as it is, for example, in the Alps, but with more rapid and prolonged convergence, one plate may be obducted over the other. This is clearly seen where the Indian Plate collided with Asia along the line of the Himalayas. The Asian Plate has overridden the Indian one, so the Tibetan Plateau now consists of a double thickness of continental crust.

Although the compressive regime of continental collision is less favorable for volcanism, local extension may permit magma to rise, particularly just after the initial collision. The volcanoes of Bulgaria, Anatolia (Turkey), and Iran are located where structural irregularities in the obducted crust have favored volcanism. Even on the Tibetan Plateau, magma has found places where it could rise through the thick crust and reach the surface. More commonly, however, magmas are arrested within the crust where they form plutonic intrusions, such as those of the Alps and Himalayas.

Depression of the roots of mountain ranges to levels with elevated temperature can result in partial melting of the continental crust. Radioactive elements concentrated in these rocks can also contribute heat, but a larger effect is that of water that is released by heating of hydrous minerals in metamorphic rocks and lowers their melting temperatures. These *anatectic* magmas tend to be rich in silica, alumina, and alkalies, especially potassium, and many contain distinctive minerals, such as sillimanite, cordierite, tourmaline, andalusite, and kyanite. The Tertiary cordierite dacites and tourmaline rhyolites of Tuscany are thought to have been produced in this way during the collision of the Adriatic and Eurasian Plates.

11.5 ▲ Oceanic Hotspots

Many volcanoes are situated far from plate boundaries on stable crust of either continental or oceanic character. There are many more of these intraplate volcanoes in the oceans, possibly because the thinner oceanic crust poses less resistance to the rising magmas but also because, not being affected by erosion, they remain as constructional features for many millions of years. Most volcanoes in the interior parts of oceanic plates form long linear chains oriented in the direction of plate motion. As explained in Chapter 9, the presently active volcanoes are at the leading edge of chains of similar volcanoes that become progressively older in the direction of plate motion. Far fewer chains of this kind are present on the continents, possibly because much more thermal energy is needed to penetrate the thicker crust. Hawaii and Polynesia are good examples in the oceans; Yellowstone is a familiar example on continental crust. The active volcanoes of the Cameroons near the western coast of Africa are at the leading edge of a chain that began in the Atlantic Ocean and, with time, migrated from the ocean onto the continent.

Most hotspot-related volcanic zones have a marked linear form, such as those seen in the Hawaiian—Emperor Chain or in Polynesia, the Marquesas, Tuamotu—Gambier, Society, and Austral Islands. The age of volcanism has a clear linear correlation with the distance from the inferred location of the hotspot, and the ratio of these two factors corresponds to the velocity of motion of the plate (Fig. 11.5). The orientation of the alignment indicates the direction of drift at the time the volcanoes

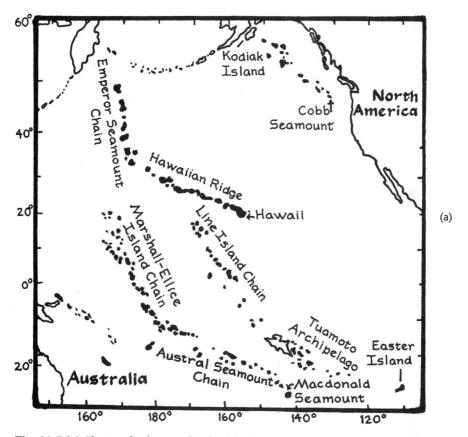

(a)

Fig. 11.5 (a) Chains of volcanic islands and seamounts in the central Pacific. The sharp bend about midway in the chains is the result of a major change in the direction of plate motion. *continued*

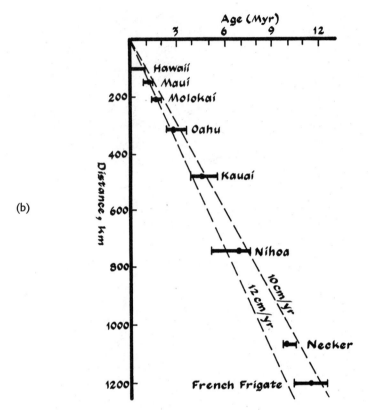

Fig. 11.5 continued (b) The relations between ages of the volcanoes of the Hawaiian–Emperor Seamount chain and distance from the hotspot just south of the island of Hawaii.

were active. It is this relation that is used to estimate the velocity and direction of plate motions.

The thermal anomalies, or hotspots, responsible for these chains of volcanoes have diameters of the order of a few hundred kilometers. In the oceans they commonly form a broad topographic swell with a diameter of as much as 1,500 km. Most hotspots remain in a fixed geodetic position as the ocean floor moves over them in the direction of plate motion, but a few seem to migrate independently of the plates at rates of less than 1 cm per year. Motion of this kind could account for the small differences of velocities of the Pacific Plate calculated from age–distance relations.

The origin of hotspots has been the subject of much discussion. Some geologists attribute them to a "thermal plume" rising from the core–mantle boundary at a depth of 2,900 km, others believe they came from a shallower source in the upper mantle. They begin with broad swelling of the crust and large outpourings of lava and then become narrower and less intense as the lithosphere drifts over them. The best example is the flood basalts of the Deccan Plateau of India, which mark the beginning of a hotspot that is now located under the island of Reunion in the Indian ocean (Fig. 11.6).

The lavas may have either alkaline or tholeiitic compositions, but the proportions differ widely from one chain to another. In Hawaii, the very earliest lavas forming Loihi, the small volcano at the leading edge of the chain, are alkaline, but the main mass of the large shields is made up of tholeiitic basalts, which, in the late, postcaldera stage, give

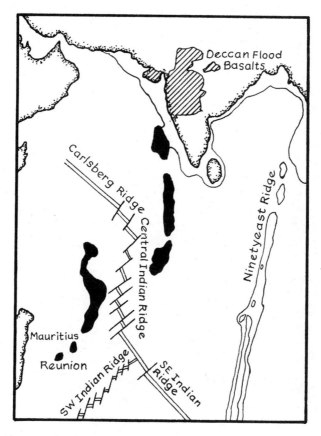

Fig. 11.6 The Reunion hotspot began with large eruptions of flood basalts in the Deccan region of India, but as the Indian plate move northward, the hotspot track entered the Indian Ocean, crossed a spreading ridge, and is now beneath the island of Reunion. (After Duncan, R. A., J. Backman, and L. Peterson. 1989. *J Volc Geoth Res* 36:193–8.)

way to progressively more alkaline lavas. The transition from tholeiitic to alkaline compositions is marked by a long interval in which the two types are interstratified. The alkaline series evolve by differentiation through intermediate compositions (hawaiites, mugearites, and benmoreites) to trachytes and more silica-deficient, alkali-rich magmas such as phonolites. These alkaline differentiates resemble those of continental rifts but have less potassium and are erupted in much smaller volumes.

11.6 ▲ Continental Hotspots

The best documented hotspot in the interior of a continent is that of Yellowstone (Fig. 11.7). Starting about 16 million years ago, a linear succession of volcanic centers has migrated at a rate of 4.5 cm per year along the Snake River Plain to Yellowstone, a distance of 800 km. Huge volumes of siliceous ignimbrites have been discharged from the large calderas of Yellowstone and Island Park (see Chapter 10), and

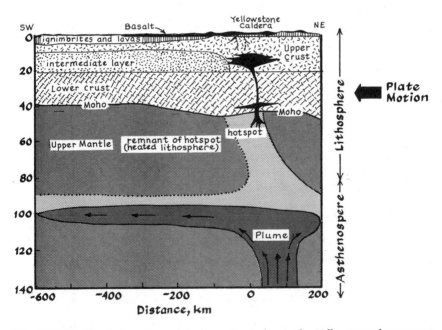

Fig. 11.7 The North American plate has migrated over the Yellowstone hotspot at a rate of 4.5 centimeters per year for the past 16 million years. It is now located below the Yellowstone caldera, but its track extends westward along the Snake River Plain. (After Smith, R. B., and L. W. Braile. 1994. *J Volc Geoth Res* 61:121–87.)

smaller amounts of basalt continue to be erupted along the entire length of the hotspot trace.

Intensive geological and geophysical studies indicate that basaltic magma rises from a fixed source in the upper mantle and collects beneath the lighter, cooler rocks of the upper crust. The overlying rocks are uplifted in a broad dome centered over the hotspot, while the influx of heat leads to partial melting of the silica-rich crust. The rhyolitic magmas generated in this way are erupted periodically as enormous ignimbrites, and the subsequent subsidence into the evacuated reservoir produces a large caldera (see Chapter 10).

As the lithosphere moves downstream, the influx of basaltic magma and heat declines, the crustal rocks cool and contract, and some of the basalt is able to ascend to the surface. This activity continues to produce small lava flows and scoria cones in the hotspots' wake along the Snake River Plain.

Suggested Reading

Basaltic Volcanism Study Project, *Basaltic Volcanism on the terrestrial planets*. 1981. Pergamon Press, 1286 p.
Chapter 6 provides an excellent overview of the relationship of volcanism to tectonic processes on the earth and earth-like planets.

Nicolas, A. 1995. *The mid-oceanic ridges*. Springer, 200 p.
An in-depth description of tectonic and magmatic processes at spreading axes as deduced from studies of ophiolite complexes.

Pritchard, H. M., T. Alabaster, N. B. W. Harris, and C. R. Neary, eds. Magmatic processes and plate tectonics. *Geol Soc Spec Publ 76*, London.
An excellent compilation of papers dealing with various aspects of the relationship of volcanism to tectonics.

Simkin, T., and L. Siebert. 1994. *Volcanoes of the world*, 2d ed. Missoula: Smithsonian Institution, Geoscience Press.
An important compilation of data on historic volcanic eruptions.

Wilson. M. 1989. *Igneous Petrogenesis, a global tectonic approach*. London: Unwin Hyman.
Contains an excellent summary of the basic relations of volcanism and tectonic processes.

Extraterrestrial Volcanism

▼

12.1 ▲ Introduction

Many of the processes and principles outlined in the foregoing chapters have been examined solely in terms of the terrestrial environment; they may not be directly applicable to other parts of our solar system where totally different conditions prevail. As space probes bring us more information about other bodies, we have had to revise many long-held, basic interpretations of volcanism based on terrestrial experience.

In the opening chapters in which we considered the origins of magmas and the ways in which they are discharged at the surface, our reasoning was based on our knowledge of the Earth's chemical and mineralogical composition, gravitational forces, temperature gradient, and distribution of heat sources, all of which are uniquely associated with the planet Earth. On small planetary bodies, much less internal heat is available for melting, the composition of the source rocks is different, and tectonic processes of the kind we associate with volcanism are either different or totally absent. The smaller the body, the weaker the gravitational forces, so magma can be ejected to greater heights. A thinner atmosphere also makes for larger eruption columns. The role played by these factors has been the subject of much recent research that has yielded a better understanding of basic magmatic processes.

12.2 ▲ Distribution of Volcanism

From the information sent back by space probes, we now know that some of the other bodies of our solar system have had even more

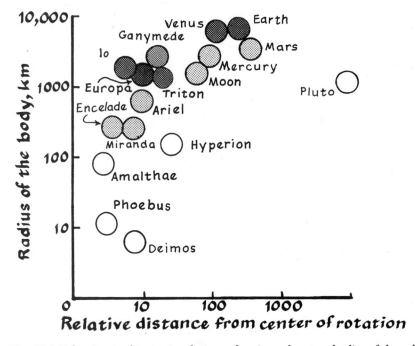

Fig. 12.1 Volcanism and tectonic relations of various planetary bodies of the solar system. Each body is represented by a circle located according to the radius of the body and its orbital radius. For example, the sun is the center of rotation of the Earth, and Jupiter that of Io. The intensity of volcanism is indicated by shading, the darkest indicating the strongest activity. (Simplified from Basilevsky, A. T. 1990. *Twelfth Soviet-American Microsymposium,* Moscow, 16–7.)

volcanism than Earth. The timing, duration, intensity, and form of this activity differ widely according to the size and position of the body (Fig. 12.1). For example, volcanism is more intense on Earth and Venus than on Mars and Mercury. A prime factor is the distance of the body from its axis of orbital rotation relative to the radius of the body about which it orbits (i.e., the sun for planets and the planets for their satellites). The effects of tidal forces are also important. Small bodies rotating close to one of large size have more intense volcanism than do large bodies orbiting at greater distances. In most cases where ages are known, volcanism was more intense in the early history of the solar system.

We now have enough information to compare volcanism on several bodies with differing sizes and situations in the solar system.

12.3 ▲ The Moon

The most familiar body, of course, is our own moon. Thanks to the Apollo exploration program, we have a wealth of information based on direct observations and samples of rock brought back to Earth by the astronauts.

It is thought that the moon resulted from an early Mars-size body hitting the Earth. The lack of volatile elements in lunar material and the overall similarity of the chemical composition to the Earth's mantle make it more likely that the body came from the Earth rather than the chance capture of a passing body.

The surface of the moon can be divided into two distinct parts: the "highlands" consisting of an irregular surface of plagioclase-rich igneous rocks, or *anorthosites*, and smooth, darker areas covered with great floods of iron-rich basaltic lavas that flowed into depressions to form the lunar maria or "seas." Despite abundant evidence of volcanism in the past, the moon is now essentially "dead." A few volcanic landforms can still be discerned, but the measured ages of all the basalts collected thus far have proved to be very old. Most are between about 3.8 and 3.2 billion years old; only a few are younger or older. Since that time, there seems to have been neither volcanism nor tectonic activity.

During a relatively brief period when the moon was still young, however, it was the scene of immense magmatism far surpassing anything known on Earth. It has been speculated that the surface was an ocean of fluid basaltic magma about 400 km deep, in which great blocks of anorthositic crust floated like icebergs in the sea.

These gigantic volumes of magma could not have been produced by normal processes of the kind that operate on the Earth; the internal heat of a body as small as the moon could not have supplied the energy needed to melt such large amounts of basalt. Although the bulk composition of the moon is similar to that of the Earth's mantle, it has lost most of its volatile components. The deficiency of these fluxing agents, together with the more rapid cooling of a body with a mass that is only about 1% of that of the earth, prevented temperatures from reaching the levels required to produce basaltic melts from the moon's own internal energy.

A more likely explanation is that the great quantities of magma were the result of especially intense bombardment by large asteroids and meteorites during the early accretionary stages of the moon's history. Although the intensity of the meteorite bombardment of the surface has since declined, it has been sufficient to reduce most of the rocks on the surface to breccia and obliterate almost all traces of the early landforms. A few circular mounds with gentle slopes resemble shield volcanoes, and the outlines of lava flows are still recognizable, but these are very subtle features. Even if the surface had not been so battered by meteorites, it would never have been one of high relief. The silica-poor, iron-rich basalts had very low viscosities, enabling them to flow for great distances on very gentle slopes. Some flow fronts are 20 to 30 m high, but this is due not to their viscosity but to the weak gravitational forces of such a small body. As we saw in Chapter 5, the force of gravity tends to make lava spread and reduces the heights of flow fronts.

During the later stages of volcanism, lavas flowed into large impact craters to form flat, crudely circular lunar maria. The smooth surface is crossed by ridges and deep, sinuous rilles that for centuries have been the subject of much speculation. The ridges have the appearance of pressure ridges not unlike those on a frozen lake, but their exact origin remains obscure. The rilles, some of which originate at craters and are large enough to be seen with a telescope, have many of the features of collapsed lava tubes, but when they were examined on the ground by astronauts, they did not have the collapsed roofs or strand lines that are common in terrestrial lava tubes (Fig. 5.5b). Another possibility is that they were produced when very hot, swiftly flowing lavas melted and eroded their substrate, but the thermal requirements for melting on this scale seem prohibitive.

Before the Apollo landings, geologists debated whether some of the lunar craters could be volcanic, but it is now clear that almost all were formed by meteorite impact. Lacking any significant amounts of water, pyroclastic eruptions of the conventional Strombolian type would be impossible. There is no doubt, however, that fountaining of the Hawaiian type (Fig. 7.1a), which requires less vesiculation, was common. Small spherules of glass blanket the surface for distances of as much as 100 km around some of the vents, and it has been estimated that because of the rapid rate of discharge and weak gravity, the jets of molten basalt reached heights vastly greater than even the most spectacular fountaining on Hawaii.

12.4 ▲ Io

Although it is about the same size as our moon, one of the satellites of Jupiter, Io, has abundant active volcanism. In fact, it is probably the most active body in our solar system. The first photographs sent back by the space probe Voyager showed nine umbrella-like eruptive plumes rising to heights of as much as 280 km. Because no impact craters are seen, recently erupted lava and pyroclastic ejecta must cover the entire surface. More than 300 active or recently active volcanoes have been identified.

The reason for the remarkable contrast with our own moon is that Io orbits the giant mass of Jupiter. The immense tidal forces of the planet, as well as those of the other satellites of Jupiter, particularly Ganymede and Europa, generate large amounts of thermal energy. It has been suggested that the interior of Io might be largely liquid.

The volcanic structures are much larger than any seen on Earth; lava flows extend for hundreds of kilometers. At least 200 calderas have diameters of more than 20 km, and an eruption plume with a diameter of about 1,400 km has been seen rising with a velocity of 500 to 1,000 m per second. There has been much speculation as to the composition of the pale grayish yellow material being erupted (see colored photograph Fig. 16). Spectroscopic analyses showed that the surface has little if any water, so the huge eruptive columns must be driven by some other volatile substance. The color suggests sulfur, but in many places, the topography is far too steep to be supported by such a weak material. The best candidate was originally thought to be a very sulfur-rich magma, but more recent studies indicate that at least some of the sulfur is superficial and that some of the magma could be a very hot, magnesium-rich basalt or peridotite.

Two types of volcanism have been distinguished on the basis of the temperatures and viscosity of the material. Small plumes are thought to be produced when red liquid sulfur at 130 to 160°C is vaporized to sulfur dioxide; large plumes could result from vaporization of sulfur that has come into contact with hot silicates. The sulfur-rich lavas are believed to come from hydrothermal deposits that have been mobilized when temperatures within the volcano rose above the melting point of sulfur. Unlike silicate melts, which become increasingly viscous on cooling, molten sulfur has a minimum viscosity at about 160°C. Above that temperature, their viscosity decreases with cooling. As a result, their flow can be quite different from that of more common lavas.

12.5 ▲ Venus

Because its size and density are similar to those of the Earth and its orbit is only slightly closer to the Sun, Venus was long thought to be a close analog of our own planet. Owing to its dense cloud cover, its surface could not be examined directly until its atmosphere could be penetrated by radar and spacecraft could land and send back information on its surface. Much of what has since been learned confirms a similarity to the Earth. The lavas, for example, resemble common tholeiitic and alkaline basalts. In other respects, however, the planet is very different.

Venus seems to lack anything resembling plate tectonics, and the scale of volcanism is feeble compared with that of the Earth. Lavas, although less frequent, flow for much greater distances than even the most fluid terrestrial flows, and although Hawaiian-type fountaining may be associated with the lavas, pyroclastic eruptions seem to have been infrequent and relatively weak.

To understand these differences, one must consider how Venus, despite its many similarities, differs from the Earth. Although its radius is only about 300 km less than that of the Earth, its mass is about 80% of the Earth's. Hence, even though the mantle contains similar amounts of radioactive elements, they generate less internal heat, and this is dissipated almost entirely by conduction through a relatively thin lithosphere.

The atmosphere of Venus is very different from our own. Apart from the lack of water, its density is greater and atmospheric pressures are nearly a hundred times greater than our own. These characteristics impede vesiculation of volatile components of magma and reduce their expansion during pyroclastic eruptions. Because Venus is closer to the sun and has a dense atmosphere rich in CO_2 that enhances the "greenhouse effect," the surface reaches temperatures as hot as 500°C. These elevated temperatures reduce the rate of heat loss from the surface of lava flows and enable them to flow much longer. At the same time, however, convection of the dense atmosphere has the opposite effect, so the overall cooling rate may not be very different from that on Earth.

While orbiting Venus, the space probe Magellan mapped almost the entire surface of the planet by means of radar images with a resolution of the order of 100 m, making it possible to distinguish relatively small geological structures. Although no eruptions have been observed, the survey revealed that Venus has had very recent volcanism that has produced thousands of shield volcanoes, domes, and calderas. As a result, much of the planet's surface has been made over to a fresh volcanic cover. If, as it appears, some of these volcanoes are still active, they could be responsible for the marked variations of the sulfur dioxide content of the atmosphere measured by the Soviet space probe Venera.

The most common type of volcano is a small shield not unlike those of Iceland. They appear to have formed from eruptions of fluid basaltic lavas from a summit vent. They are remarkably symmetrical with a diameter of a few kilometers and a height of only 100 to 200 m. Somewhat larger features with flat tops and steep edges have more in

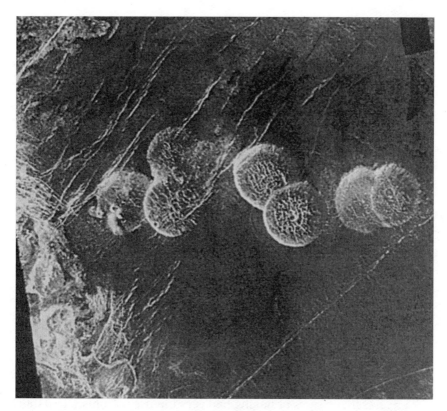

Fig. 12.2 Radar image of Venus taken by the space probe Magellan in 1991 shows seven domes of lava on the edge of Alpha Regio. Their average diameter is about 25 km. (Photograph by NASA.)

common with domes. A notable example is the chain of seven domes on the eastern margin of Alpha Regio (Fig. 12.2). Measuring 25 km in diameter and up to 750 m in height, their forms resemble those of the silicic domes of the Coso and Mono Lake fields of California.

Some of the larger volcanoes have much in common with the great shield volcanoes of Hawaii. They have diameters of a few hundred kilometers, heights of up to 5 km, and steep-sided summit calderas. The cone of Sif Mons is a typical example. It has a diameter of 300 km and a height of 1700m. Lava flows 200 to 300 km long and 15 to 30 km wide radiate from the summit. A rift zone is marked by chains of craters and grabens, and a caldera about 45 km in diameter occupies the summit.

Several large caldera-like depressions have been formed on relatively level topography with no apparent relation to large central-vent volcanoes. These gigantic structures, which measure 100 to 200 km across and 1 or 2 km in depth, are difficult to interpret. In some ways they resemble giant calderas or impact craters, but they lack steep rims. Instead, they are bound by quasi-circular rings of faults and grabens with eruptive vents that have poured lava into the interior depression. One theory holds that they are the result of subsurface drainage of magma. This could explain the sagging centers and tensional faulting around the margins, but there is no obvious place where the magma could have gone.

12.6 ▲ Mars

In contrast to Venus, Mars occupies an orbit with a radius 50% greater than the Earth's and thus receives less solar radiation. The mechanisms of eruption of ancient martian volcanoes are poorly understood, but they must have differed from those of terrestrial volcanoes because of the conditions that prevailed on that planet. Its atmosphere is only 1% as dense as our own. Hence, the surface temperatures are so low that even CO_2 freezes and lava flows are cooled more quickly. Pyroclastic eruptions, on the other hand, encounter less impedance from the atmosphere and can reach great heights. This effect has been enhanced by the smaller size of Mars, which, with a radius of only 3,390 km, has a gravitational force only one-third that of the Earth's.

Mars has many signs of having had very intense volcanism less than a billion years ago but now seems to be nearly extinct. The planet is divided into two more or less equal parts separated along a line inclined 30 degrees from the equator. The southern hemisphere consists of ancient terrain that has been cratered by countless meteorites and possibly comets. In contrast, great sheets of lava cover two extensive areas in the northern hemisphere, Elysium Planitia and Tharsis Montes. The huge dome of Tharsis, with a diameter of 8,000 km and elevation of about 10 km, covers a quarter of the surface of the planet. Its volcanism began about 3.8 billion years ago and may have ended during the eruptive phase of Olympus Mons 400 to 100 million years ago. Statistically, the martian volcanoes appear to be less numerous but correspondingly larger than those of the Earth because they have had more time to grow in a setting with an immobile lithosphere and no plate tectonics.

Scattered about the northern region are numerous isolated volcanoes that, like the great shields of Hawaii or Reunion, have thick piles of lava and summit calderas. Three types of morphologically distinct volcanoes are found in close association. Among the oldest volcanic features are the "patera" (literally "broad bowl"). These have a very

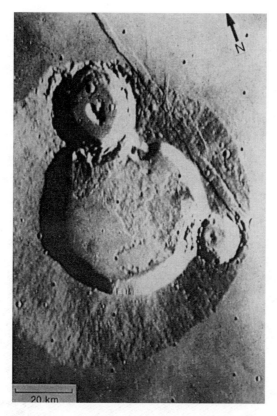

Fig. 12.3 Ulysses Patera, a volcano on Mars photographed by the space probe Viking in 1976. The volcanic crater (the largest structure in the center of the photo) is modified by two meteorite impacts recognizable from their central peaks and cut by a graben running about north–south. (NASA photograph and Lunar and Planetary Institute.)

low profile and slopes of less than one degree. Alba Patera, for example, is 1,200 km in diameter and about 7 km high. Another example, Ulysses Patera, is illustrated in Fig. 12.3. Some of the paterae have shallow, irregular summit calderas and are thought to be formed by ignimbrites rather than by lava flows.

A second type, known as "tholi" (cupola-like structures), is not sharply distinguished from the first but is smaller and tends to have steeper slopes. Although some appear to have had relatively recent explosive eruptions, their main periods of volcanism seem to have been during the earliest history of the planet. Many appear to have been partly buried under later lavas that flooded the surrounding lowlands.

The third type, called "mons," includes the great shield volcanoes, such as Olympus Mons (Fig. 12.4). This formidable edifice rises to a height of about 27 km, making it the largest mountain in the solar system. With a diameter of 500 km, the shield is surrounded by a larger polygonal, nearly circular outline governed in part by the structure of the rocks on which it stands. The gentle slopes of 4 to 6 degrees have concentric terraces resulting from radial compression caused by the gravitational forces on the huge mass. Numerous flows more than 100 km long and 25 to 80 m thick are piled one atop the other. An oval summit caldera measuring 80 by 90 km is made up of at least six large coalescing and crudely aligned depressions. A gigantic talus slope 2 to 8 km high seems to be the result of large-scale slope failures and land-

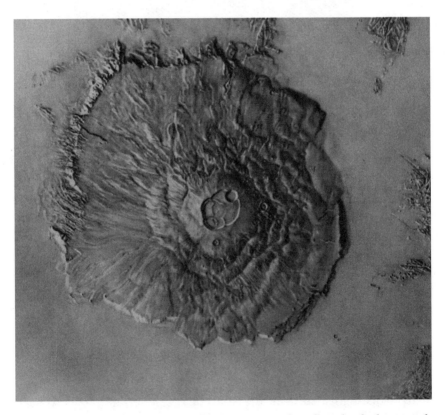

Fig. 12.4 The giant shield volcano Olympus Mons is thought to be the largest volcano in the solar system. It is morphologically very similar to the shields of Hawaii and Réunion but vastly larger. The caldera resembles that of Mauna Loa but measures 60 km across—about 20 times the diameter of its Hawaiian counterpart. (Photograph by NASA.)

slides on the scarps surrounding the volcano. A large aureole extending 700 to 1,000 km beyond this talus may have been caused by ash or ignimbrites, but it could also be older eroded lavas or landslide debris.

12.7 ▲ Other Bodies in the Solar System

Although we have less information on other planets and their satellites, our knowledge continues to expand. Spacecraft visiting Mercury have shown that its surface consists entirely of volcanic rocks, but the activity seems to have been concentrated in the earliest stages of its history with a brief renewal of activity about 3.8 billion years ago. Earth-based observations at radar wavelengths recently showed that permanently shadowed craters at the poles of Mercury may have deposits of ice, possibly brought in by comets. They resemble the ice deposits that some believe to be at the polar regions of the Earth's moon.

Although we see intense volcanism on Io, the other satellites of very large planets seem to be very different, both in the nature of their activity and the composition of their eruptive products. Two satellites of Jupiter—Callisto and Ganymede—as well as those of Saturn—Encelade and Rhea—had volcanoes of a very special kind. The latter are composed of a form of ice rich in the ammonia that was an important constituent of these bodies. The volcanoes were formed about 3.5 billion years ago when internal temperatures were above −100°C, the melting point of ice with this composition. Triton, a satellite of Neptune, seems to have geysers produced by vaporing large amounts of nitrogen. Photographs of Miranda, a satellite of Uranus, taken by Voyager in 1986 show traces of activity that is even stranger. It is difficult to say whether it is of volcanic origin or not.

As future studies of the other planets continue to reveal strange features such as these, we will gain a better understanding of the processes underlying volcanism on our own planet.

Suggested Reading

Basaltic Volcanism Study Project. 1981. *Basaltic volcanism on the terrestrial planets.* New York: Pergamon Press, 1286 p.
Chapters 4 and 5 review the nature of basalts on the planets that had been studied up until 1981.

Fielder, G., and L. Wilson. 1975. *Volcanoes of the earth, moon, and mars.* London: St. Martin's Press.
Although somewhat outdated, the book contains an excellent description of the different forms that volcanism takes under the varied conditions on the planets and moons of our solar system.

Francis, P. W. 1993. *Volcanoes: A planetary perspective.* New York: Oxford University Press, 443 p.
An introductory text containing an extensive section dealing with volcanism on other bodies in the solar system.

Smith, D. K. 1996. Comparison of the shapes and sizes of seafloor volcanoes on Earth and "pancake" domes on Venus. *J Volc Geoth Res* 47:47–64.
A study of the physical factors governing the morphology of large volcanoes on two different planets.

Hydrothermal Phenomena and Geothermal Energy

▼

13.1 ▲ Hydrothermal Processes

Most large volcanoes have hot springs and fumaroles around their flanks. These thermal manifestations are said to be "hydrothermal" because hot water or steam is always their dominant component. Although most is meteoric ground water, it normally has a certain proportion that is of magmatic origin. Its overall composition has little resemblance, however, to the gases and dissolved elements described in Chapter 3. Not only are these magmatic components diluted, but their character changes as they lose heat and react with the rocks through which they move.

Volatile emissions take on a much different character when their temperatures approach those at which water vapor begins to condense. Owing to the large amount of heat involved in condensation of water, temperatures tend to be buffered in the narrow range in which steam and liquid water coexist. At the same time, the proportions of the other gases begin to change rapidly. Because CO_2 has a much lower solubility in hot liquid water than H_2S, the former is lost in gas emissions, while relatively more of the latter is retained. For this reason, the proportions of CO_2 and other high-temperature components, such as SO_2, H, Cl, and F, become less important with declining temperatures and increasing distance for their source. In general, gases discharged at high temperatures are rich in CO_2, whereas emissions at temperatures at or below the boiling point contain relatively more H_2S.

All hydrothermal processes entail a transfer of heat and dissolved matter in an aqueous or gaseous medium. Their manifestations have

213

(a)

Fig. 13.1 (a) Strokkur, one of the Icelandic geysers from which the name has been taken, erupts every 10 minutes or so and reaches a height of about 15 m. This photo shows it at the moment that a mass of superheated water is emerging from the vent and beginning to vaporize. (**a,** Photograph by J. M. Bardintzeff.)

a variety of forms and compositions. *Fumaroles* and *solfataras* are synonymous terms for thermal areas in which temperatures are close to or above the boiling point of water and the quantities of water vapor and other gases are large compared with the amount of liquid water. Vents that emit CO_2-rich gases with little water vapor are known as *mofettes*. This is in contrast to hot or warm springs, in which the volume of steam is small compared with that of liquid water, and the main gas is usually H_2S. Springs are by far the most numerous and widespread type of hydrothermal manifestation. *Geysers* (from the Icelandic word for "spout") are rare but often spectacular hot-water eruptions in which steam and boiling water are periodically expelled in jets or sprays (Fig. 13.1).

Although hydrothermal manifestations are common in craters, even of long-dormant volcanoes, hot springs are much less common on the slopes than around the base. This is because rain water and melted snow percolate down through the highly permeable ash and lava and emerge only at lower elevations. Hot springs are typical of basins where the water table is shallow and abundant groundwater circulates through thick accumulations of permeable sediments, tephra, and fractured rock (Fig. 13.2). They are by no means restricted

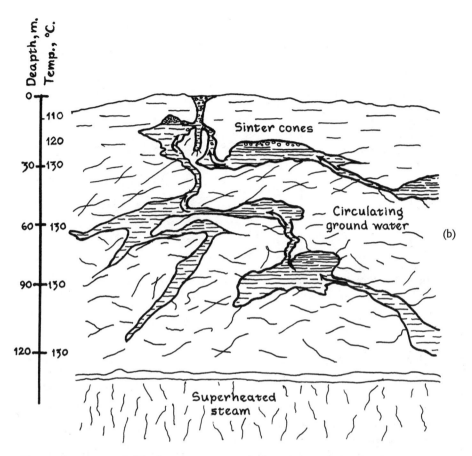

(b)

Fig. 13.1 continued (b) The eruptions result from heating of groundwater in a network of channels a few hundred feet below the surface. As the water reaches its boiling temperature, it is converted to steam that collects in cavities and begins to drive the liquid water upward. As the hot water rises the relief of pressure enables it to vaporize, which in turn accelerates the flow so that the entire column of water flashes to steam. The expelled water is then replaced by an influx of fresh groundwater, and the process is repeated. The temperature gradient shown on the left is based on measurements in a borehole near Old Faithful. (**b**, After Barth, T. F. W. 1950. *Carnegie Instit Wash Pub* 587:174.)

to localities of recent volcanism but are also found in many regions where older rocks have been subjected to strong deformation or where abnormally steep temperature gradients result from intrusions of magma.

Geysers are among the rarest but most renowned geological phenomena on earth. All but a few are in three regions of recent

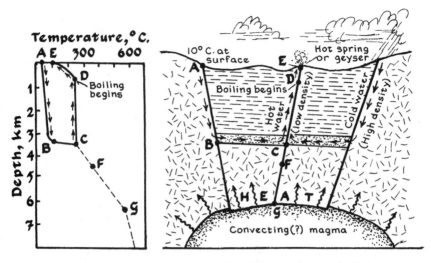

Fig. 13.2 Simplified schematic diagram of a convective, high-temperature geothermal system. (Modified from White, D. E. 1968. *U. S. Geol Surv Prof Paper 458-C*, p. 109.)

volcanism. Yellowstone National Park has the greatest number—about 200; Iceland has about 30 active geysers; and New Zealand, 20. The largest geyser on record was that of Waimangu, New Zealand, which from 1890 to 1940, erupted every few hours, sending 800,000 liters of hot water to heights of up to 460 m. For well over a century, Old Faithful in Yellowstone has continued to erupt every 60 to 90 minutes. The eruptions are caused by the combined action of heat and pressure on shallow groundwater. The water in large cavities is raised to temperatures as hot as 260°C in an unusually steep geothermal gradient but is prevented from boiling by the pressure of the overlying water column. As its temperature rises and its density is decreased, the hot water rises to shallow levels of lower pressure and begins to boil and expand more rapidly. This sets off a chain reaction by lowering the density and hence the pressure of the water column, inducing further boiling, and sending the entire column to the surface. Cooler water then flows into the subterranean reservoir, and the cycle is repeated.

The times required for meteoric water to circulate through hydrothermal systems varies between wide limits depending on the dimensions and volume of the reservoir and the rate of emission. In the case of the Valles Caldera of New Mexico, for example, it is estimated to be as long as 2,000 years. During this time, the compositions of waters are modified by the rocks through which they flow, and the proportions of dissolved components gradually change according to the nature of the rocks. At the volcano Arenal in Costa Rica, magmatic components in hot springs on the flank of the active cone account for 68 to 92% of the dissolved substances. NaCl, KCl, and Na_2SO_4 are dissolved from fumarolic incrustations on fractures, HCl and H_2SO_4 are absorbed from volcanic gases, and Ca and Mg are derived from alteration of mafic volcanic rocks. The amount of material carried in solution varies according to the nature of the eruptive activity and the degree of prior alteration of the rocks. Springs around the base of an inactive volcano may discharge moderate amounts of dissolved material, such as calcium carbonate, but those associated with active volcanoes generally carry much greater amounts, mainly in the form of chlorides and carbonates of sodium with lesser amounts of calcium (see Table 13.1).

A wide variety of solids are deposited as precipitates from hot springs or as sublimates from gases and water vapor. The Yellowstone thermal area is justly famous for its colorful deposits in and around the hot springs, fumaroles, and geysers. These deposits can be related to three general types of thermal waters: (1) those containing large amounts of calcium carbonate, (2) acidic waters rich in sulfates, and (3) alkaline waters with large concentrations of chlorides. The compositions of some typical examples are given in Table 13.1.

TABLE 13.1

Typical compositions of thermal waters
(amounts are reported in parts per million)

	1	2	3	4
pH	2.1	9.0	2.2	9.0
Fe	168	6	13	tr.
Mg	60	tr.	3	—
Ca	132	3	13	4
Na	—	70	25	372
K	—	—	50	31
SO_4	1603	17	474	21
Cl	5	30	45	435
F	—	—	—	22
CO_3	0	16	—	100
S	—	0	—	0
H	—	—	tr.	0
NH_4	—	—	107	—
B_2O_3	—	—	—	18

1. Acid water from a hot spring in Iceland, from a compilation by Barth (1950).
2. Alkaline water from a spring near no. 1, also from Barth (1950).
3. Acid–sulfate water from mud pot in Norris Basin, Yellowstone National Park (From Allen and Day, 1935, p. 107, no.1).
4. Alkaline water of Old Faithful Geyser, Yellowstone National Park (From Allen and Day, 1935, p. 249)

Carbonate-rich waters are confined to areas underlain by limestone or dolomite. The Mammoth Hot Springs of Yellowstone are a good example. The springs, which rise through thick beds of limestone, deposit beautiful terraces of carbonate-rich *travertine* when the water reaches the surface and exsolves CO_2 (Fig. 13.3a). The loss of CO_2 and acidic gases raises the pH of the water and thereby reduces the solubility of carbonates until they begin to precipitate on the margins of the pools.

The acid–sulfate and alkaline–chloride types of springs differ in their hydrologic relationship to the water table. Acidic springs tend to have smaller discharges with larger amounts of steam. They form shallow, turbid pools and boiling mud pots (Fig. 13.3b) and are more common on slopes than in basins. Their low pH results primarily from oxidation of H_2S to H_2SO_4 in a separate vapor phase, but in springs and fumaroles that are closely associated with direct volcanic emanations, HCl may also be important. The waters are rich in volatile components, such as B, CO_2, H_2S, and NH_3, and contain various amounts of Mg, Fe, Ca, Na, and Si, depending on the degree to which these elements are leached from the surrounding rocks. Volatile metals, such as Hg, Pb, Bi, Sn, and As, are not uncommon.

Alkaline–chloride types have lower concentrations of volatile components but are rich in Na, K, Cl, SO_4, HCO_3, and CO_3, all of which have high solubilities in liquid water. In contrast to the acid–sulfate waters that result mainly from condensation of steam above the water table, the alkaline–chloride type is dominated by liquid water that has circulated through permeable rocks at elevated temperatures.

As far back in history as we can see, thermal springs have been praised for their therapeutic value. From the fifth century BC, when Hippocrates praised those on the island of Kos, down to modern times when Franklin Roosevelt made the warm springs of Arkansas famous for treatment of polio, thermal waters have been credited with special curative powers. Most springs of this kind result from meteoric water infiltrating through hot crustal rocks, where they gain heat and soluble mineral components before returning to the surface decades or even centuries later. The waters are enriched in various cations, such as Ca, Mg, Na, K; a variety of metals; and anions, such as HCO_3^-, SO_4^{2-}, Cl, and F. They may also contain dissolved gases derived from deeply buried magmas.

13.2 ▲ Mineral Deposits

Thermal waters are responsible for many of the world's richest mineral deposits. Processes of mineral deposition have been observed

Fig. 13.3 (a) Deposits of calcareous sinter around fumaroles and hot springs at Yellowstone. **(b)** A boiling mud pot of the kind found in hydrothermal regions throughout the world.

directly in fumaroles in the rhyolite erupted in 1912 in the Valley of Ten Thousand Smokes, Alaska. Chlorine- and fluorine-rich gases coming from the hot interior of the pumice flows produced incrustations of hematite; sulfides of Cu, Pb, and Zn; and various compounds of Mo, Co, Ni, Sn, Mn, Ti, Sb, Bi, B, As, Se, and Te. Wells drilled for steam in the Salton Sea region of southern California discharge chloride-rich brines with 332,000 ppm dissolved solids, including large amounts of Na, K, Ca, Li, Fe, Mn, Ag, B, Ba, Cr, Cu, Ni, Pb, Sr, and Zn. A sample of the material deposited on discharge pipes contained silver and gold concentrations of 381 and 0.11 ounces per ton, respectively. Copper in the form of the mineral bornite amounted to 20% of the deposit. Although a number of youthful rhyolitic domes testify to recent volcanism within the structural graben, isotopic ratios of hydrogen and oxygen indicate that the steam has little if any magmatic component. Instead, meteoric water falling in the Salton Sea basin circulates to deep levels, where it gains heat and dissolved components from the thick sedimentary section filling the depression.

Most of the mineral deposits exploited for metals, such as copper, lead, zinc, uranium, tin, molybdenum, mercury, and even gold and silver, are products of ancient hydrothermal systems. Meteoric and magmatic fluids circulating through the crust transfer these elements in solution and deposit them in concentrations that can be mined economically.

13.3 ▲ Geothermal Energy

Geothermal systems can also be a potential source of energy. It has been estimated that with present technology, they could supply as much as 10% of the United States' energy needs, and as techniques for extracting it are improved, it could yield much more. Geothermal energy has the distinct advantage that it does not contribute pollutants to the atmosphere. Plants powered by bituminous coal produce about 5.5 metric tons of carbon dioxide per day for every megawatt of electricity. In contrast, a typical geothermally driven plant releases about 0.15 metric tons.

The first commercially successful geothermal power plant was developed at Lardarello, Tuscany, in 1827 when the Count of Lardarel undertook the first drilling to exploit the Earth's heat. His pioneering efforts were responsible for what is today a major source of Italy's energy. The project produces 412 megawatts. In recent years, natural steam has been developed in many regions of recent volcanism, such as Iceland, New Zealand, California, Japan, and Central America. Early development was centered around areas of conspicuous hy-

Fig. 13.4 The geothermal plants at The Geysers, California, produces an amount of energy equivalent to that consumed by the nearby city of San Francisco. (Photograph by Julie Donnelly-Nolon, U.S. Geological Survey.)

drothermal activity, but as boreholes are drilled to greater depth, it has been found that the energy resources can be much greater than surface manifestations indicated. At The Geysers in northern California, for example, only a few fumaroles and hot springs were associated with the nearby Clear Lake volcanic field, but with drilling extending to depths of about 3,000 m, enough steam is produced to make this the largest producer of geothermal energy in the world. It now has a capacity of more than 2,000 megawatts of power, enough for the entire needs of the city of San Francisco (Fig. 13.4).

Most large geothermal fields are related to shallow bodies of magma. The amount of magma erupted at the surface is only a small part, perhaps 10%, of the total that has risen into the crust. This means that as much as 90% of the magma in a recently active volcanic region is stored below the surface.

When meteoric water descends to depths of anomalously steep geothermal gradients, it draws on the large amount of heat stored in the rocks, and when tapped by wells and brought to the surface as steam, it can drive turbines for generating electrical power (Fig. 13.5). The economic feasibility of such systems is largely a function of the amount of heat carried in a given amount of water and the temperature levels at accessible depths. At temperatures greater than 150°C, the *enthalpy,* or heat content of the water, is great enough to produce

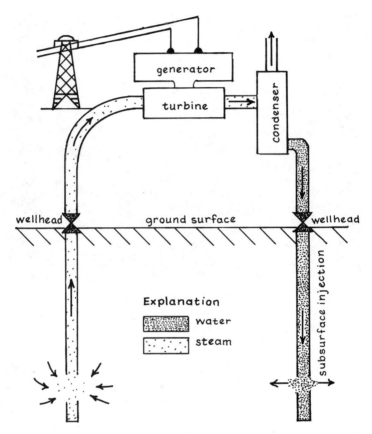

Fig. 13.5 Schematic diagram showing how natural steam is drawn from a bore-hole and fed through turbines to produce electrical energy. After extracting its heat, the condensed water is returned to the subsurface reservoir at injection wells. (After Duffield, W. A., J. H. Sass, and M. L. Sorey. 1994. *U. S. Geol Surv Circ* 1125: 63 p.)

"dry" superheated steam, but in moderate-to-low enthalpy systems, wells produce a mixture of liquid water and steam with a much lower thermal efficiency. After using the steam, the remaining hot water is normally injected back into the ground so that it can be recycled through the heat source.

Even in the absence of steam, however, it is possible to recover the energy in hot water by putting it through heat exchangers and using certain easily vaporized fluids, such as isobutane, to drive the turbines. The hot water can also be used to heat buildings. Most of the homes in Klamath Falls, Oregon, are heated in this way. In Iceland, hot water pumped from shallow wells is used in hot houses to grow a variety of fruits and vegetables, including even tropical fruits such as bananas.

Iceland produces 935 megawatts or about 5% of its national requirements from natural steam. Italy, which was the first country to develop this source of power on a large scale, now produces 548 megawatts. New Zealand began to exploit geothermal power about 50 years ago and now produces about 340 megawatts from two large geothermal fields. In California, about 1,000 megawatts are produced from at least 700 wells in a dozen different localities. Geothermal energy is also important in Japan, Philippines, and Central America. On a global scale, however, this is not a major energy source. All together, natural steam accounts for less than a percent of the world's total energy production because it can be exploited only in countries that have the requisite geological conditions, but it can be of great importance in volcanic regions where other forms of energy, such as coal and oil, have to be imported. Naturally, the feasibility of geothermal energy depends on a number of technical and economic factors, such as the volume and area of the thermal anomaly, the permeability of the rocks, the balance between heat losses and recharge, as well as hazards, accessibility, investment costs, and the use to which the energy will be put.

Suggested Reading

Bowen. R. 1989. *Geothermal resources*, 2d ed. London: Elsevier.
Includes information on all the important geothermal localities of the world.

Duffield, W. A., J. H. Sass, and M. L. Sorey. 1994. Tapping the earth's natural heat, *U S Geol. Surv Circular* 1125: 63 p.
An excellent summary of the sources and economic potential of geothermal energy.

Fisher, R. V., G. Heiken, and J. B. Hulen. 1997. *Volcanoes, crucibles of change.* Princeton, NJ: Princeton University Press.
Chapters 11, 12, and 13 give an excellent review of the hydrothermal resources, mineralization, and other beneficial aspects of volcanism.

Nicholson, K. 1993. *Geothermal fluids: Chemistry and exploration techniques.* Berlin: Springer-Verlag.
A comprehensive work on mineralization and other geothermal phenomena.

Volcanoes and the Human Environment

▼

14.1 ▲ Volcanoes and Natural Catastrophes

Humans have always had an equivocal attitude toward natural disasters: fear of the incomprehensible, anger when faced with the inevitable, and hope of ultimately overcoming the elements of nature.

In recent years, the number of victims of natural catastrophes has been growing at a regular rate of about 6% per year, or roughly three times the rate of demographic growth. As more and more of the Earth's population is concentrated in cities that continue to spread at an uncontrolled rate, the populace is exposed to increasing risk. This problem has raised such concern that an agency of the United Nations devoted the last decade of the 20th century to gaining a better understanding of natural catastrophes, providing information to citizens at risk and, so far as possible, reducing the impact of disasters. A permanent office in Geneva is charged with the responsibility of coordinating disaster relief.

Although volcanic eruptions cause fewer deaths than other natural events, such as hurricanes and earthquakes, the numbers are large and continue to grow. UNESCO has identified about 100 volcanoes throughout the world that pose substantial risks. The great majority of these are in densely populated countries in the Circum-Pacific "Circle of Fire."

Reliable statistics on the impact of eruptions have been available only since about 1700 (Table 14.1). They show that since that time, 27 eruptions have resulted in at least 1,000 deaths, or a total of more than a quarter of a million victims in all. During this same period, all other eruptions combined were responsible for a total of about 10,000 deaths. Thus, it appears that large volcanic eruptions are relatively

TABLE 14.1

Volcanic catastrophes since 1700 that were responsible for at least 1,000 deaths

Volcano	Country	Year	Pyroclastic Flows	Lahars*	Tsunamis	Famine	Gas
Awu	Indonesia	1701		3,000			
Oshima	Japan	1741			1,475		
Cotopaxi	Ecuador	1741		1,000	1,000		
Makian	Indonesia	1760		2,000			
Papandayan	Indonesia	1772		2,957			
Laki	Iceland	1783				9,336	
Asama	Japan	1783	1,151				
Unzen	Japan	1792			15,188		
Mayon	Philippines	1814	1,200				
Tambora	Indonesia	1815	12,000			80,000	
Galunggumg	Indonesia	1822		4,000			
Mayon	Philippines	1825		1,500			
Awu	Indonesia	1856		3,000			
Cotopaxi	Ecuador	1877		1,000			
Krakatau	Indonesia	1883			36,417		
Awu	Indonesia	1892		1,532			
Soufriere	St. Vincent	1902	1,565				
Mt. Pelee	Martinique	1902	29,000				
Santa Maria	Guatemala	1902	6,000				
Taal	Philippines	1911	1,332				
Kelut	Indonesia	1919		5,110			
Merapi	Indonesia	1930	1,300				
Lamington	New Guinea	1951	2,942				
Agung	Indonesia	1963	1,900				
El Chichon	Mexico	1982	1,700				
Nevada del Ruiz	Columbia	1985		25,000			
Nyos	Cameroon	1986					1,746
Total victims			60,090	50,099	54,080	89,336	1,746

Lahars include victims of associated pyroclastic flows that cannot be distinguished from posteruption lahars due to rains.

rare events but very deadly. On average, two catastrophic eruptions take place each century. During the 20th century, the 1985 eruption of Nevado del Ruiz in Columbia and the 1902 eruption of Mt. Pelée in Martinique resulted in 25,000 and 29,000 deaths, respectively. During the 19th century, Indonesia was hit especially hard with the 1883 eruption of Krakatao causing 36,000 deaths and the 1815 eruption of Tambora 92,000.

If these same eruptions were to occur today, the toll would be even greater. The destruction of Pompeii and Herculaneum by the AD 79 eruption of Vesuvius (Fig. 0.3) took about 16,000 lives, but this was a relatively minor event compared with what would happen to the pre-

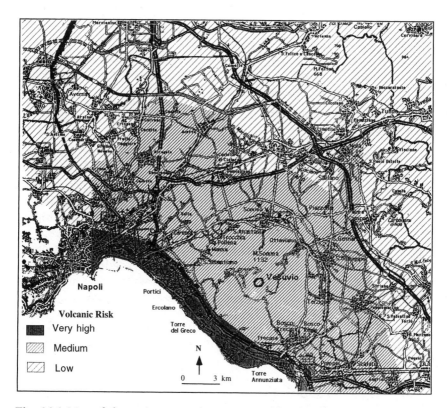

Fig. 14.1 Map of the region around Vesuvius and Naples showing the degree of risk. Note that much of Naples and the towns concentrated along the shore of the Bay of Naples would suffer the greatest impact if a major eruption, such as that of AD 79 were to occur today. (After Scandone, R., G. Arganese, and F. Galdi. 1993. *J Volc Geoth Res* 58:261–73.)

sent population around the volcano if a similar eruption were to occur today (Fig. 14.1). The loss of life would be at least as great, and the loss of property and economic disruption would be enormous. It is estimated that more people would be killed in the rush to evacuate than from direct effects of the eruption.

14.2 ▲ Seven Types of Volcanic Hazards

It is convenient to distinguish primary risks directly related to a volcano from secondary risks resulting from other consequences of an eruption, such as mudflows (lahars), fire, and famine.

Just as the church has defined seven deadly sins, volcanologists have defined seven deadly types of volcanic hazards: lava flows, ash falls and ballistic ejecta, pyroclastic flows, gases, lahars, avalanches, and tsunamis. The first four are the main primary hazards, the remaining three are secondary effects.

Lava Flows

We noted in Chapter 5 that on reaching the lower flanks of a volcano, lavas tend to be channeled into valleys. They move slowly, normally less than a few kilometers per hour. These lavas constitute no real danger to human lives because there is usually time to avoid them, but what can one do about buildings and other structures standing in their way? Attempts to divert lavas with artificial dams and levees or explosives have had only limited success. During the 1973 eruption of Helgfell on the island of Heimaey near the coast of Iceland, a lava flow was successfully retarded by pumping sea water on it at rates of up to 12,000 tons per hour. Attempts to stop the advance of lavas by constraining them with dams rarely succeed, but it may be possible to divert a flow into a new channel. In 1983, for example, volcanologists succeeded in diverting a lava flow from Etna by dynamiting a break in the upper parts of its levee and causing it to take a new course.

Other attempts to divert lava flows were made during an eruption of Etna that lasted 473 days between December 14, 1991, and March 30, 1993. More than 250 million m³ of lava covered an area of about 7 km² at an average rate of 6 m³ per second. On January 2, 1992, when a flow with a thickness of 10 m had descended 5.5 km from an eruptive fissure on the flank, the army and fire fighters undertook to construct a barrier. After 10 days of uninterrupted work, the structure measured 234 m in length and 21 m in height and contained 370,000 m³ of soil and rock. At that time, the lava was only 150 m away, but on March 14th, it reached the base of the barrier and began to pond behind it. It passed around the southern end on April 7th, and on the 8th, overtopped the crest (Fig. 14.2). When it reached a steeper slope, the lava advanced more rapidly, covering another kilometer in five days and threatening the town of Zafferana Etnea and its 7,000 inhabitants. Six successive attempts to construct earthen barriers proved fruitless. Beginning on April 13th, explosives were placed in the levee well up the slope in an unsuccessful attempt to tap the molten interior and redirect the flow. On the 14th of April, the lava overflowed the last barrier, destroying two outlying houses of Zafferana and covering nearby orchards. By the following day, it had descended to an elevation of 750 m, 7.5 km from its source. The lobe

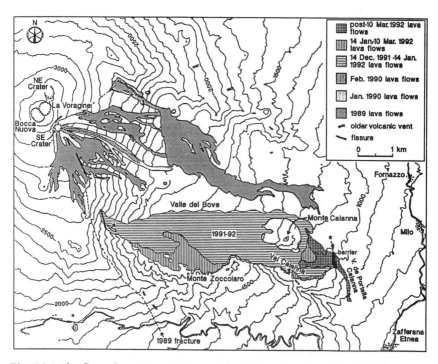

Fig. 14.2 The flows from Etna in December 1991 and much of the following year descended eastward from a source at an elevation of 2,200 to 2,400 m following the Valle del Bove until they came within 2 km of Zafferana Etnea. The map shows the barrier that was constructed between the 2nd and 12th of January and overtopped on April 8th. (After Romano, R., et al. 1992. *Bull Global Volc Network* 17:4.)

threatening the town then ceased to advance, and the inhabitants returned to their homes. From then until May 6th, various attempts were made to breach the lava tubes. Explosives were set off in the levee, and large blocks of concrete were dropped into the flow from helicopters, but the effects were inconclusive. Finally, on the 27th of May, a lava tube was successfully disrupted by means of 7 tons of explosives, and the flow took a new course. In the end, the main lesson was that it is enormously difficult to dam the front of an advancing flow, but under certain favorable conditions, it may be possible to divert it into a new channel.

The path that lava will follow is easy to anticipate if one takes into account its viscosity and the relief of the threatened region. Laboratory simulations carried out after the eruption of Etna succeeded in modeling the flow with remarkable accuracy. Computer-based modeling has been used successfully in Japan (Fig. 5.2). Using techniques such

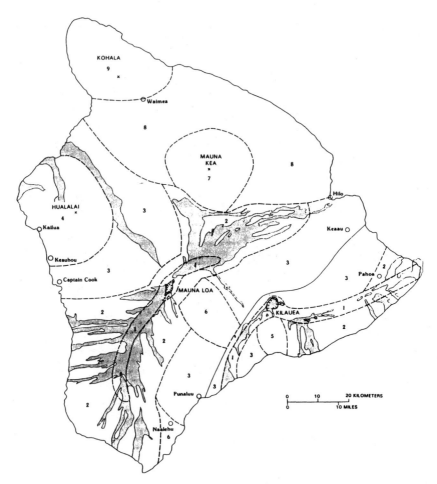

Fig. 14.3 Map of the island of Hawaii showing zones of relative risk from lava flows. A solid line separates the structures of Mauna Loa and Kilauea, and shading indicates areas covered by historic lava flows up to 1975. Dashed lines outline nine zones that are numbered from 1 to 9 in decreasing order of risk. Zone 1, which includes the summits and the active rifts of Mauna Loa and Kilauea, has the maximum risk; 25% of its surface is covered by flows in the last two centuries. Zone 9 consists of the volcano Kohala, which has not erupted for 60,000 years but cannot be considered extinct. (After Mullineaux, D. R., D. W. Peterson, and D. R. Crandell. 1987. *U. S. Geol Surv Prof Paper* 1350:599–621.)

as these, one can prepare maps showing zones that are vulnerable to various degrees of risk. An example for the island of Hawaii (Fig. 14.3) shows how areas of high risk have been delineated around Mauna Loa and Kilauea.

Ash Fall

Of all the hazardous products of volcanism, ash fall has the most widespread effect. A layer of lapilli can reach thicknesses of several meters and cover entire towns. Fine dust has a harmful effect on the respiratory system, but the effects on humans are usually brief and easily mitigated. The effect on vegetation can be more serious; thick ash falls can damage vegetation, leading to crop failures and death of livestock. It can damage various types of machinery, including aircraft engines. When carried into streams, ash accelerates erosion and can bring on floods and mudflows.

Although no two eruptions are exactly the same, the distribution of earlier ash falls can serve as a general guide to future eruptions of the same volcano. Their regional extent is a function not only of the volume of pyroclastic material erupted but also of the degree of fragmentation and the manner in which they are dispersed (see Chapter 7). Using meteorological records, one can predict the probable wind patterns at different elevations. With this information, risks can be anticipated and structures designed to alleviate their effect.

Pyroclastic Flows

All types of pyroclastic flows pose serious risks. Unlike ash falls, which tend to be dispersed over great areas, most pyroclastic flows follow drainage lines and are normally confined to a single sector radiating outward from the source of the initial explosion. The energy may be concentrated in a narrow angle, so its effect is much greater than if it were dissipated over a wide radius. Morphological features, such as the form of a dome and the slope of the volcano, control the trajectory and velocity of the flow. Pyroclastic flows can, of course, surmount topographical barriers, as they did in the 1991 eruption of Mt. Unzen, Japan. Prior warning of this hazard is crucial because the impact is too sudden to permit evacuation once a flow has begun.

A large ignimbrite eruption of the kind that covered thousands of square kilometers around the calderas of Yellowstone and Long Valley (see Chapters 8 and 10) has never struck a populated region in historical times. Sooner or later, however, such an eruption will occur, and when it does, one can only hope that there will be adequate warning to permit the large-scale evacuation that would be the only possible response to such a threat.

During the recent period of more or less continuous unrest at Long Valley caldera in southern California, geologists recognized that a disasterous eruption could threaten the nearby communities of Mammouth Lake and Bishop, but they met with strong resistance on the part of merchants and real estate agents who sought to minimize the danger and discredit warnings.

Gases

Because toxic volcanic gases are widely dispersed and difficult to escape, they constitute a major hazard. For example, great amounts of SO_2 emitted from Laki in Iceland in 1783 reacted with associated water vapor in the atmosphere to produce sulfuric acid that had a devastating effect on crops. Fluorine can be a dangerous pollutant in the hydrological system.

Carbon dioxide is especially dangerous because it is invisible and odorless. Heavier than air, it forms a layer close to the ground following valleys and collecting in depressions. Humans and livestock are unaware of the danger and are overcome by asphyxiation. During the 1783 eruption of Laki, great numbers of sheep were killed when they sought refuge in depressions where gas had collected. On February 20, 1979, 0.1 km^3 of almost pure CO_2 was discharged during a phreatic eruption on the Dieng Plateau of Java, killing 142 persons.

The volcanoes of the Cameroons pose an unusually severe risk of this kind. The water in the lakes filling old volcanic craters becomes saturated with carbon dioxide, which becomes unstable and is suddenly released in huge quantities. Gas eruptions of this kind have occurred periodically for centuries.

The eruption of Lake Nyos in 1986 was the latest and most dramatic of this kind. Situated near the village of Wum, the elliptical lake measures 1,800 m across its greater axis. The flat bottom is 208 m below a surface that is maintained at a constant level by a natural outlet. The sudden emission of a flood of CO_2 on the 21st of August killed 1,746 humans and wiped out most of the livestock. Although trees were knocked down in a few places, the bodies of the victims showed few external signs of injury. Although chemical or thermal effects were not evident, concentrations of 10 to 20% CO_2 in the air seem to have been fatal. Many of the 845 survivors suffered damage to their lungs and eyes.

The distribution of dead animals showed that the concentration of lethal gas reached a level 120 m above the lake before a cloud of heavy gas with a volume of about 1 km^3 and 15% CO_2 flowed 25 km down valleys and into topographic depressions (Fig. 14.4). After the gas was released, the level of the lake dropped about a meter, and a reddish stain was observed near the center. A similar event took place on December 30, 1986.

Three different hypotheses have been postulated to explain this combination of events:

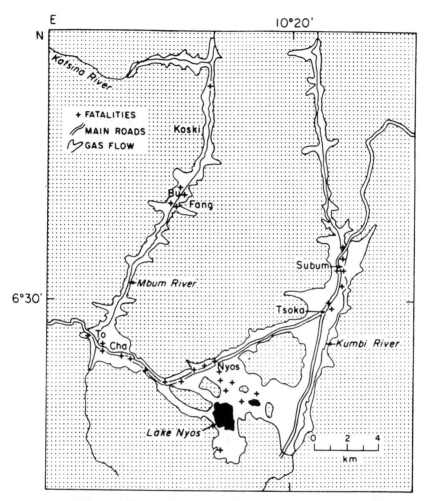

Fig. 14.4 Areas affected by the deadly gas that came from Lake Nyos on August 21, 1986. (After Morin, S. and J. Pahai. 1987. *Rev Géogr Cameroun* 6:81; Sigurdsson, H. 1987. *EOS* 68, 23:570–3.)

1. A bacterial process in the lake water resulted in decomposition of organic material and a slow accumulation of CO_2 that eventually became unstable.
2. A large amount of volcanic gas was suddenly injected from beneath the lake.
3. Dissolved gas and bicarbonates had accumulated in the water over time; a seismic shock disturbed the equilibrium of a stratified column of gas-saturated water and triggered exsolution and liberation of CO_2.

The last hypothesis seems most plausible; it would explain the reddish stain as iron precipitated by the degassing. It calls for a preliminary stage during which magmatic CO_2 slowly accumulated in the water and a sudden release of gas when the equilibrium was disturbed.

Conditions near the lake remain dangerous because the water still contains an estimated 265 million m^3 of dissolved CO_2 to which another 5 million is added each year. The hazard is complicated by the natural dike controlling the level of the lake. If this dam should fail, it could allow a sheet of CO_2–charged water to flow as far as Nigeria, 108 km away. Careful studies of other lakes of the Cameroons have found that not all are dangerous. For example, Lake Bambuluwe, being only 58 m deep, has too little capacity to store important amounts of gas. The most dangerous lakes are situated along a tectonically controlled fissure system, whereas the safer ones have no visible relation to active volcanism.

In central France, the maar of Sénèze, about 100 km southeast of Clermont Ferrand, may have had eruptions 1.5 million years ago that were similar to that of Lake Nyos. The evidence is found in unusual concentrations of animal remains—elephants, bears, oxen, and horses—that were suddenly killed, most likely by asphyxiation.

Theoretically, it should be possible to take measures to mitigate future catastrophes such as these by pumping out gas-charged water near the bottom and thereby reducing the amount of gas that could be liberated at the surface. An experiment in which water was siphoned from deep in another Cameroon lake, Monoun, showed how this could be done. A vertical pipe 90 m long and about 5 cm in diameter was lowered into the lake, and after pumping to start the flow, the rise of water became self-sustaining by virtue of the expansion of the exsolving gas. The gas–water mixture flowed at rates of up to 151 liters per second. With three pipes 14 cm in diameter, the 10 million m^3 of CO_2 could be removed from Lake Monoun in less than a year. In April of 1995, a pipe 205 m long and 14 cm in diameter was placed vertically in Lake Nyos. Because 1 liter of the deep water can contain up to 17 liters of CO_2, the mixture of water and gas rises at an accelerating rate and fountains to heights of 20 m at velocities of about 100 km per hour. The system could continue for years until all the gas is removed. It might even be possible to put the gas to commercial use.

Lahars

Lahars or mudflows result from a combination of special circumstances in which loose volcanic material is mobilized by water (see Chapter 8). The name *lahar* comes from Indonesia, where the phenomenon is a familiar hazard (see Table 14.1).

The most recent catastrophe of this kind, which took place in Columbia on the 13th of November 1985, took 25,000 lives. The volcano, Nevado del Ruiz, had a complex history going back 1.8 million years with 12 important eruptive events in the course of the last 11,000 years, but the local population viewed it as peaceful and referred to it as the Leon Dormido (Sleeping Lion).

Seismic activity was noted during November of 1984, about a year before the paroxysmal eruption. The following February, emissions from fumaroles increased; on September 11th, a rain of ash lasted six hours, and blocks up to a meter in size were hurled for distances of up to a kilometer. In October and November, local and foreign volcanologists foresaw an increased risk of lahars in a region that had already been indicated on maps of volcanic hazards. Their concern was well founded. A relatively mild Plinian eruption on November 13th produced four important lahars that were triggered by melting part of the summit glaciers and suddenly releasing meltwater from pockets under the ice. Sixteen percent of the surface area of the ice cap and snowfield (4.2 km^2), and 9% of the volume (0.06 km^3) was melted.

At 10:30 in the evening, the lahars reached the town of Chinchina 60 km to the west and buried it along with 2,000 inhabitants. By 11:35 PM, more lahars came 60 to 80 km down the eastern and northeastern slopes, reaching the towns of Armero (23,000 victims) and Mariquita (Fig. 14.5). In the Azufrado River, the average speed of the lahars was 10 m per second or 72 km in two hours, and their average flow rate was 25,000 to 30,000 m^3 per second. The maximum flow was 48,000 m^3 s^{-1}, and near the source the wave reached a height of 40 m. At the level of Armero, the flow was still 2 to 5 m thick and moved at a rate of 8 m s^{-1}. The momentum was enough to destroy the town.

Although the hazard was clearly recognized and volcanologists urged an evacuation when the volcano showed signs of unrest, the response was delayed—with tragic results. Since the eruption, a permanent volcanological observatory has been established at Manizales.

During the series of major eruptions of the Philippine volcano Pinatubo in 1991–1992 and for several years thereafter, numerous lahars were generated during each rainy season (see Chapter 8). An alert system was set up by the Philippine Volcanological Institute so that observers at points high on the volcano could warn the inhabitants of towns and villages that a lahar was descending. This allowed half to three-quarters of an hour for evacuation. In addition, a system of barriers was constructed. Although these measures reduced the loss of life, the material damage was still great.

The lahars at Nevado del Ruiz resulted from melting snow; those of Pinatubo were brought on by heavy rainfall. Others have been triggered by a sudden release of water from crater lakes. In 1919, 38 million m^3

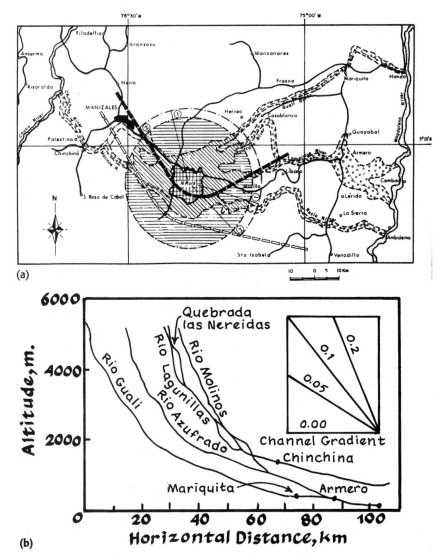

Fig. 14.5 (a) Map of risk zones around Nevado del Ruiz as they were delineated on November 12, 1985, one day before the eruption. Five types of risk are identified: lava flows (grid pattern), lahars (pattern of small open circles), directed blasts (sectors of high risk marked *1* or moderate risk marked *2*), and ash fall (high risk in the area north of the heavy broken line and moderate north of the lighter dashed line). Areas of high risk from pyroclastic flows are shown by diagonal ruling, and those of moderate risk by horizontal ruling. **(b)** Longitudinal profiles of the valleys down which lahars flowed from Nevado del Ruiz. Towns that were affected are indicated on their respective valleys. (**a,** After Cepeda, H., et al. 1985. *Memoria explicativa* 1–28; **b,** after Pierson, T. C., et al. 1990. *J Volc Geoth Res* 41:17–66.)

of acidic lake water were suddenly released from the crater of Kelut in Indonesia, triggering lahars that took 5,110 lives (see Table 14.1). After the eruption, an even larger lake formed, threatening a catastrophe at any moment. In 1926, the Dutch government, which administered Indonesia at that time, conceived a bold plan to bore lateral tunnels through the walls of the lake and drain the water. The level was lowered by 56 m, and the volume was reduced from 65 to 3 million m³. As a result, the eruption of 1951 produced no casualties, but the tunnels became obstructed, and in 1966 hundreds more were killed. A new tunnel constructed by the Indonesians in 1967 has subsequently maintained the lake at a safer level.

As mentioned briefly in Chapter 8, prehistoric lahars have descended from Mt. Rainier, Washington, and flowed over much of what is now the heavily populated region near Puget Sound. One must assume that they will recur in the future, but a well-designed warning system, if coupled with a plan for rapid evacuation, could greatly alleviate the risk to humans. During the period that the Costa Rican volcano Irazu was erupting, in 1963 through 1965, a well-organized system of this kind proved very effective in averting loss of life when lahars destroyed 15 city blocks of the city of Cartago. The Costa Rican system, excellent as it was, was relatively simple compared with the one required for evacuating a city the size of Tacoma.

Landslides or Avalanches

Landslides on volcanoes can be triggered by inflation of the volcano and by steepening of the slopes, as they were at Mt. St. Helens in 1980 (see Chapter 8). In a more general sense, any steep-sided volcano may have unstable slopes. This is true, for example, of many of the volcanoes of the Cascades. Volcanoes tend to be less stable than other types of mountains of equivalent size because their fracture systems and ash layers provide planes of weakness along which the slopes can fail. An eruption is not needed to trigger avalanches; they often follow earthquakes or periods of prolonged rain or melting snow. Tropical terranes are especially vulnerable because of the deep weathering and heavy rain.

In assessing risks, the main factors to be considered are the steepness and relief of the slope, its orientation, the lithologic nature and coherence of the material, and the climate and vegetation. As in the case of pyroclastic flows, the extent of an endangered region is a function of the slope angle and elevation of the source. Landslides can block valleys damming the drainage and creating potentially unstable lakes. Six events of this kind have been recorded since the beginning of the 20th century, and an estimated 20,000 persons have died from this cause in the last 400 years.

Tsunamis

An explosive eruption near a coast line, either on land or under water, can cause abrupt slumping of sediments on steep slopes. This, in turn, can trigger waves of great amplitude, which strike coastal areas with enormous destructive power. Tsunamis, as these waves are called, can cross the entire width of an ocean. The coast of Chile, for example, has been hit by tsunamis generated in Japan. In deep water, the amplitude of the wave may be so small that it escapes notice, but its height increases rapidly as the depth of water decreases closer to the shore.

Before it erupted on August 27, 1883, the volcano Krakatao was a small island in the Sunda Straights between Java and Sumatra. Waves generated during the eruption struck the nearby coasts of Java and Sumatra killing thousands of persons and carrying away entire villages. Of the total death toll of 36,417, almost all were from the tsunamis. The waves generated by the eruption crossed the Pacific and Indian Oceans and were detected by tide gauges along the coast of France 17,000 km from their source. It appears that the tsunami was triggered when an avalanche, similar to the one that occurred on the slope of Mt. St. Helens in 1980, entered the sea.

14.3 ▲ Hazard Evaluation

Risk has been defined as the product of three factors: value, vulnerability, and hazard. *Value* is the potential impact measured in terms of the number of lives and the value of property at risk; *vulnerability* is the percentage of lives or property likely to be lost; and *hazard* is the probability of a certain area being affected by a particular volcanic phenomenon. Risk considered in this way is useful to distinguish cataclysmic eruptions, which fortunately occur only once or twice each century, from more frequent but less serious events, such as eruptions of Kilauea, which take place every few years without causing serious damage. Between these two extremes, there is a broad range of intermediate cases that are more difficult to evaluate. By estimating the probability that an important eruption will occur and determining the nature of a particular hazard, one can estimate the potential risk for a populated region. This risk includes the chances of loss of life or property or of economic disruption in a threatened zone.

The notion of risk can be viewed in terms of several factors: first, the eruption and its direct consequences; second, its subsidiary effects, such as lahars and tsunamis; and, third, the effects on the nearby populace and threatened structures, such as buildings, dams, or bridges. Effects of this third type include famine, fires, and floods caused by ruptured dams. The civil administration must consider all of these potential hazards and set up emergency plans for evacuation, sanitation, food, and shelter.

Among the most difficult effects to prepare for are those that modify the ecological equilibrium. For example, the eruption of the Guatemalan volcano Santa Maria in 1902 caused the death of hundreds of thousands of birds, and without their predators, mosquitos proliferated in great numbers leading to an epidemic of malaria that caused more deaths than the direct effects of the eruption.

It is difficult to calculate the total cost of a volcanic eruption. As a general indication, the costs of three moderate eruptions in Japan have been estimated at 18 million dollars for the eruption of Izu-Oshima (November 15 through 22, 1986), 190 million dollars for that of Miyake-jima (October 3 through 4, 1983), and 210 million dollars for Usu (August 6 to October 27, 1978). For the 1991 eruption of Pinatubo, the cost is estimated at 708 million dollars for the first two years and as much as $1 billion for the entire period from 1991 to 2005.

The volcanologist is faced with a difficult challenge when he or she attempts to evaluate risks of this kind. Because every volcano has its own special character, one must know its history, just as a doctor must know the history of a patient. One can study prehistoric products to determine the pattern of eruptions and the timing of previous events, but the record is seldom complete, even in the most thoroughly studied volcanoes. Some of the methods used to date past eruptions have been discussed in earlier chapters. Charcoal buried under pyroclastic debris and protected from the atmosphere can be dated by the ^{14}C method and thereby provide a record for the last 50,000 years. Going farther back in time, other radiometric methods make it possible to determine the age of lavas that are hundreds of thousands of years old.

Studies of this kind have been carried out for several of the large volcanoes of the Cascade Range. They show that past eruptions have been separated by long intervals of repose and that some of these periods of inactivity have been longer than the time that has elapsed since the last recorded eruption. Thus, the lack of historical activity cannot be taken as evidence that a particular volcano is extinct. On the contrary, the largest eruptions of the past have usually followed the longest periods of repose.

Given an adequate record of past eruptions, a statistical pattern can be extrapolated into the future. Many volcanoes have a certain rate of production that, despite irregularities in the frequency of eruptions, is nearly constant over periods of centuries (Fig. 4.3). Since its eruption in AD 79, the average production of magma from Vesuvius has been 1.5

to 2 million m³ per year. Assuming the system has not changed since the last eruption in 1944, it appears that there is presently a volume of about 40 to 70 million m³ of "available" magma ready to erupt. A single discharge of this magnitude would result in the most important eruption since the sub-Plinian event of 1631.

To prepare for risks such as this, one can compile maps (Fig. 14.5a) showing the distribution of potential hazards in the region around a volcano. Maps of this kind integrate several factors, such as the possible extent and thickness of ash falls and pyroclastic flows, lahars, and areas that have been affected to differing degrees by previous eruptions of the volcano.

14.4 ▲ Surveillance by Volcanic Observatories

An observatory constructed in the vicinity of a volcano to maintain systematic surveillance is the ideal method of monitoring its behavior and forecasting eruptions. A permanent staff of scientists can record visible events and a variety of geophysical and geochemical observations that, when measured periodically, provide signs of changing conditions. Today, only a few volcanoes are monitored in this way. The U.S. Geological Survey maintains three: one on the Hawaiian volcano Kilauea, another near Mt. St. Helens in the Cascade Range, and a third in Anchorage, Alaska. These observatories use a variety of techniques to monitor potentially active volcanoes.

Seismic Activity

Seismometers are by far the most practical tool for monitoring a volcano. Several instruments at appropriate positions on the flanks of a volcano can be used to detect the slightest disturbance. A type of seismicity, usually referred to as "harmonic tremor," is characteristic of subsurface movements of magma. It normally comes in closely spaced harmonic waves lasting one or two minutes. Although very small in magnitude, they can overwhelm the recording capacity of a nearby seismograph. This type of seismicity is particularly common when magma is rising through an eruptive conduit and as such provides a useful means of forecasting an impending eruption. Seismographs on Kilauea in Hawaii commonly record tremors of this kind a few days or hours before magma breaks out from a shallow reservoir. On a larger scale, many eruptions occur within a day or two of large, nearby earthquakes. For example, two months before the 1980 eruption of Mt. St. Helens, an earthquake of magnitude 4.2 was recorded at a shallow depth below the volcano, and another of magnitude 5.2 was registered at a depth of 1.5 km at the same time as the slope failure that triggered the eruption at 8:32 AM on May 18.

Ground Deformation

Regular measurements of the levels of precisely determined bench- marks at selected points on the slope of a volcano can detect tilting of the surface and provide valuable information on the inflation or defla- tion of an underlying magma reservoir. Several methods are used to obtain integrated measurements. (1) Tiltmeters of the Blum type mea- sure the torque exerted by displacement of a pendulum on a horizon- tal axis and convert this mechanical force to an electrical signal indi- cating the position of the pendulum. From this, one can determine changes of slopes as small as a few microradians. The signal is trans- mitted to the observatory and recorded at regular intervals. (2) Precise measurements of the distance between two fixed reference points by means of laser geodometers with microreflectors can detect variations of about a few millimeters over distances of several kilometers. On a more local scale, extension across dilational fractures is measured by extensometers. These measurements can detect inflation caused by a rise of new magma into shallow levels of the volcano and the deflation that follows its discharge at the surface. Every year the Icelandic fis- sure volcano, Krafla, goes through an average of three cycles of infla- tion and deflation. (3) Precise measurements of topographical eleva- tions at selected benchmarks, if carried out at regular intervals over a period of years, can reveal the cumulative deformation of the volcano. At Kilauea, this deformation results from changes of the volume of magma stored at depths of about 9 km. Horizontal or vertical dis- placement at the surface can be as great as half a meter over a broad area of 500 km². Measurements of this kind of deformation are greatly facilitated by the Global Positioning System.

Deformation of volcanic structures has been modeled numerically, assuming uniformly isotropic elastic rocks. Most calculations of this kind show that a reservoir of magma lies at a depth of a few kilome- ters below the summit. It may not be a single large body but rather a network of interconnected dikes and sills.

Variations of Magnetism, Gravity, and Electrical Properties

Continuous measurements of the intensity of the magnetic field of Piton de la Fournaise, Reunion, have shown changes of as much as 10 nanoteslas (nT). Individual excursions last from less than a minute to several months. They appear to be directly correlated with volcano- seismic crises; it has been observed that certain eruptions have been preceded a few days earlier by variations of as much as 5 nT. The observations are directly linked to changes of the remnant or induced magnetism of the rocks, the electrical resistivity of the ground, and circulation of ground water.

Similarly, monitoring variations of the gravity field and the electrical potential of the ground help define eruptive cycles. As new magma rises into the volcano and inflates the overlying structure, the redistribution of mass is easily detected by measuring changes of the gravity field, and the change of temperature alters the electrical conductivity of the rocks.

Temperatures and Chemical Composition of Water and Gases

Monitoring changes of temperature and the chemical composition of fumaroles and crater lakes or meteoric water percolating through the volcanic rocks can often detect variations within a cycle of activity (see Chapter 3). A good example is the Japanese volcano Usu, which has been monitored systematically for nearly half a century. It has been found that following formation of the dome of Showa-Shinzan in 1943 to 1945, degassing of the magma resulted in long-term changes in the proportions of Cl, F, CO_2, and S and an increase in the amount of water vapor as more meteoric water penetrated the cooling dome.

Gases rising through the ground on and around volcanoes have been found to be reliable indicators of activity. The radioactive gas radon-222 (^{222}Rn) is a minor but very useful magmatic component. The product of decay of radium-226, it has a half-life of only 3.8 days. Because the ^{222}Rn being produced by minute amounts of radium defuses rapidly through the magma, rocks, or soil, a new intrusion of magma at shallow depths below a volcano quickly produces a halo of ^{222}Rn that is easily detected by instruments at the surface. Thus, by monitoring the ^{222}Rn in the vicinity of a potentially active volcano, one can detect a new influx of magma even before it erupts. This technique has been especially useful in monitoring the Nicaragua volcano Cerro Negro.

Surveillance by Artificial Satellites

Satellites that constantly circle our planet can furnish valuable information about the state of volcanoes. Multispectral images, especially those in the infrared range, can record thermal variations in and around a volcano. *Argos* marker beacons situated on or near a volcano transmit information by way of satellites directly to the observatories monitoring the volcano. Using a network of these markers with the

Global Positioning System (GPS), one can measure deformation of the Earth's crust on an almost continual basis. On July 13, 1989, an eruption of the volcano Teishi a few kilometers offshore from the Izu-Oshima Peninsula of Japan was closely monitored in this way. The relative displacement between two stations measured before and after opening of an eruptive fissure reached 14.5 cm ± 3 mm horizontally and 5 cm ± 10 mm in elevation.

Compiling Records of Past Activity

The basic approach to the problem of forecasting and mitigating the effects of eruptions is to install instrumentation and obtain information that can aid in predicting future behavior. One of the most serious problems is that of monitoring potentially active volcanoes that have had no recent eruptions. A list of 80 dangerous volcanoes published by UNESCO in 1984 did not even mention Nevado del Ruiz, the Columbian volcano that only a year later was responsible for 25,000 deaths. Unless one can track the physical and chemical characteristics of volcanoes during successive periods of repose, it is difficult to recognize any change of conditions that could possibly be threatening.

An understanding of the mechanisms of large eruptions poses special problems because the amount of factual information we have is still too sparse for precise interpretations. At this stage, only 5% of the Quaternary volcanoes of the world are properly studied. As long as this condition continues, the next cataclysmic eruption could take place where it is least expected.

14.5 ▲ Civil Defense

In order to reduce the level of panic at the time of a catastrophe, it is important to educate the threatened population by explaining exactly what a volcano is capable of doing, the various dangers it poses, and what one must or must not do in case of an eruption. In some countries, notices are distributed with detailed instructions, usually in the form of drawings: listen for instructions on the radio; assemble essential supplies, such as drinking water and extra clothing; do not go to the schools to get children because they will be taken care of there; and do not make unnecessary use of the telephone. Japan, a country that is particularly concerned with natural hazards, is well organized in this respect. The city of Kagoshima situated at the foot of the volcano Sakurajima was devastated on January 12, 1914. Since then, every year on the anniversary of the catastrophe, the population of 500,000 inhabitants practices an evacuation using all available means of transportation.

One of the best examples of the effectiveness of education and planning was the evacuation of Rabaul in Papua New Guinea. The town is situated on the rim of a caldera, one side of which is open to the sea. In 1937, after several days of earthquakes and uplift, a violent eruption caused 500 deaths, mainly from pyroclastic flows. When a new period of earthquakes and uplift began in 1971, the population had grown much larger and the potential for disaster was obvious. When the activity increased during 1983 and 1984, an intense educational campaign was organized. The geologists at the Rabaul Volcano Observatory set up a warning system with four stages of alert, ranging from slight activity that called for no immediate response to Stage 4 when eruptions could be expected within days or hours. At this highest level of alert, when thousands of earthquakes were felt each day, the authorities would order an evacuation according to a carefully prepared and rehearsed plan. A few years later, this plan was put to the test. As renewed seismic activity gradually increased over a period of years, the level of alert was raised accordingly until 1994 when it became apparent that an eruption was imminent. The plans for Stage 4 were invoked, and the population was evacuated in an orderly fashion 20 hours before the outbreak of an eruption. The well-organized plan has been credited with averting a disaster of major proportions.

The different services for civil protection must compile information that can be used to study and mitigate volcanic crises. They must also coordinate agencies responsible for rescue, evacuation, supplies of food and water, and medical and social assistance while taking into account the many possible courses an eruption might take. A supply of safe drinking water must be assured, because pollution of the normal water supply can quickly bring on an epidemic. Special medical services should be ready to deal with grave burns of the skin and respiratory system, edema of the eyes and lungs, infected wounds, and obstruction of the respiratory and digestive systems by ash.

14.6 ▲ Global Effects: Volcanism and Climate

Volcanoes inject large quantities of material into the atmosphere in the form of solid particles, gas, and aerosols. These products may reach the lower limit of the stratosphere (or tropopause at elevations between 10 km in the temperate atmosphere and 17 km in the tropics). In some cases, they rise even higher. The prehistoric eruption of Taupo, New Zealand, for example, probably sent a column as high as 50 km (see Chapter 7). At least 28 eruptions since that of Krakatao in 1883 had eruption columns that reached the stratosphere where the ash was entrained in the jet stream. The ash cloud erupted by Krakatao in 1883, for example, circled the earth for three years. Aerosols composed of microscopic particles of liquid suspended in gas are formed by a photochemical reaction of sulfur gases and water vapor in the stratosphere. They create a haze that can absorb or reflect part of the solar radiation while absorbing infrared radiation emitted from the base of the atmosphere.

On a more local scale, ash-laden eruptive clouds can be a hazard for aviation. Between 1978 and 1991, 23 incidents were recorded in which aircraft encountered volcanic ash clouds. During the eruption of Galunggung in 1982 and 1983, two Boeing 747s lost power in all four engines and had to descend without power from an elevation of about 8,000 m before the pilots were able to start the engines again and land the plane in a normal fashion. Similarly, five passenger planes were damaged during the eruption of Redoubt Volcano in Alaska between December 1989 and February 1990. In one case, a Boeing 747 that was descending for a landing at Anchorage lost power in all four engines. It continued to descend without power from 8,534 to 4,267 m before power was regained at an elevation that was only 1,219 m above the surrounding mountain peaks. Fortunately, the pilot was able to land safely.

During the eruption of Pinatubo in 1991, 16 commercial aircraft came into contact with the ash cloud. One of these was 1,200 km from the volcano. The engines of six of the planes were seriously damaged, but no lives were lost. Better communication between volcanologists and air-traffic controllers is needed to avoid such accidents in the future.

Components held in suspension at high elevations can have a marked influence on climate. Benjamin Franklin, while serving as the first U.S. ambassador to France, was the first to see a connection between Europe's unusually severe winter of 1783 through 1784 and the eruption earlier that year of the Icelandic volcano Laki. Exactly 100 years later, measurements made after the eruption of Krakatao showed a 10% decrease of solar radiation and a substantial 0.4°C drop of the average temperature throughout most of the globe for a period of three years.

The Mexican volcano El Chichon erupted between March 28 and April 4, 1982, sending ash to heights of 17 to 31km. During the following months, the average worldwide temperature fell a quarter of a degree (Fig. 14.6). The climatic effect was reflected in the growth rate of trees, which preserve a record of temperature variations in the thickness of their annual growth rings.

Aerosols filter the sun's rays more than ash particles suspended in the high atmosphere. Sulfuric acid plays a major role, as do chlorine

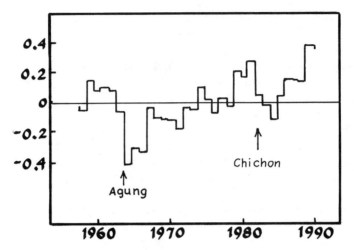

Fig. 14.6 Variations of global temperatures in the troposphere between 1958 and 1989. The annual values are expressed as deviations in degrees C. The effects of the eruptions of Agung (1963) and El Chichon (1982) are clearly apparent. These two eruptions discharged only modest volumes of solid material (<0.5 km³ each) but produced large amounts of aerosols. (After Angell, J. K. 1990. *Geophys Res Ltrs* 17.8:1093–96.)

and fluorine. We have noted that some fissure eruptions, such as those of Laki and Eldgja in Iceland, were accompanied by strong discharges of gas that had a greater effect on climate than some large explosive eruptions. The amount of volatiles emitted can be estimated after the fact by measuring the amount of acid in layers of ice laid down in polar glaciers or by analyzing the volatile content of fluid inclusions trapped in the eruptive material.

Four historic eruptions produced between 10^{10} and 10^{11} kg of SO_2, H_2S, Cl, and F that had a marked effect on climate. In order of decreasing importance, these were Tambora (1815), Laki (1783), Eldgja (934), and Katmai (1912). Comparisons of temperatures before and after the eruptions have shown an average cooling of 0.2 to 0.5°C for one to five years.

Eruptive plumes from the Philippine volcano Pinatubo during 1991 reached heights of as much as 40 km where they had a notable effect on the stratosphere. After the eruption, information on conditions in the atmosphere (the "Aerosol Optical Thickness [AOT]") was obtained directly by the NOAA 11 satellite (Fig. 14.7), and in August of the same year, similar information was obtained for the eruption of Hudson Volcano in Chile. The eruption cloud from Pinatubo completed its first circuit of the earth on July 7th and then continued to make additional

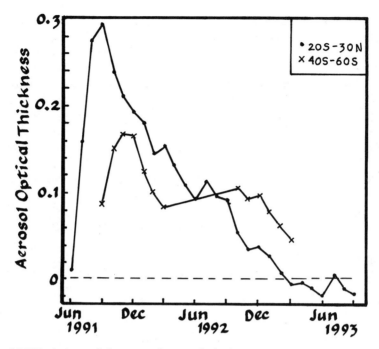

Fig. 14.7 Variations of the Aerosol Optical Thickness as a result of particles suspended in the atmosphere following two eruptions in 1991. The measurements were made at two belts of latitude. The peak in the belt at 20°S to 30°N resulted from the eruption of Pinatubo in 1991, whereas the peak that was measured slightly later between 40°S and 60°S was from the eruption of Hudson volcano in Chile. (After Stowe, L. 1993. *Bull Global Volc Network* 18:9.)

circuits between 30°N and 20°S, where the AOT coefficient reached a maximum between August and September. Then, starting in October, the same effect was seen between 40°S and 60°S, in large part as a result of the activity of Hudson.

Have large volcanic eruptions in the more-distant past brought on "volcanic winters"? Could they have caused extinction of certain forms of life, and could they do so in the future? A mysterious cloud covered Europe, the Middle East, and Asia in AD 536 and 537. The sun at its zenith is thought to have had one-tenth of its normal strength. This effect may have been linked to a volcanic eruption in the southern hemisphere, such as that of Rabaul, New Guinea, that has been dated at the year 540 ± 90. Fourteen million years ago, the eruption of the Roza flow, one of the Columbia River flood basalts in Washington and Oregon, produced more than 10^4 million tons of aerosols. The proportion of the sun's radiation transmitted through this pall would have been roughly the amount of light received from a full moon.

The role of volcanism is often mentioned in the continuing debate over the cause of mass extinctions at the end of the Cretaceous period 66 million years ago. While many scientists attribute the event to the impact of a comet or meteorite, others cite the emission of great quantities of solid and gaseous volcanic material during eruption of the Deccan flood basalts of India, possibly at the same time as similar eruptions in the North Atlantic. The two phenomena could have been related. A statistical study of impact craters, periods of mass extinction, and discharge of large volumes of flood basalts shows a clear correlation and a periodicity of 28 to 32 million years over the last 600 million years. This interval corresponds fairly well to the half–wave length of the oscillations of the suns angle with respect to the plane of our galaxy. With each passage through this plane, the earth is bombarded by comets. Calculations show that large impacts have enough energy to affect the Earth's mantle and could initiate large-scale volcanism.

14.7 ▲ Useful Contributions of Volcanoes

The Earth's active volcanoes demonstrate that it is a living planet very different from a dead body like the moon. The great quantities of gas contributed to the primitive atmosphere by almost continuous volcanic activity during the early history of the Earth gradually enriched it in nitrogen, carbon dioxide, argon, and water vapor. Most of the water was condensed and added to the oceans, and volcanic emanations under the sea modified the composition of seawater to create the conditions essential for the appearance of the first forms of life.

Throughout geological time, volcanism has played a central role in forming the environment in which humans developed and flourished. It has contributed many of the essential elements of life, including most of the water of the seas and the gases of the air we breathe. The components of terrestrial rocks, including most sedimentary and metamorphic rocks, if traced back to their origins, were products of magmatic processes.

Soil Fertility

Since the human race first evolved in the shadow of the volcanoes of the East African Rift, some of the densest concentration of population on earth have been in the close vicinity of volcanoes. This is especially noticeable in the tropical latitudes. For example, the island of Java has many active volcanoes and one of the densest populations in the world, whereas the adjacent nonvolcanic islands of Borneo and the Celebes are among the most sparsely populated. The explanation for this difference lies in the greater fertility of tropical soils that receive a periodic influx of volcanic ash.

Because the products of volcanoes are normally rich in certain components, such as Ca, Mg, K, Na, P, and S, they act as a natural fertilizer making it possible to cultivate crops in tropical regions where the soil would otherwise lack the nutrients essential to plants. A combination of warm temperatures and heavy rainfall leads to rapid depletion of soils, so most tropical lands can be farmed profitably for only two or three years after the natural forest is removed. Unless fertilizer is used to replenish components that are lost when the delicate balance of the natural system is upset, productivity declines rapidly. Periodic ash falls from volcanoes can offset this depletion and turn otherwise barren lands into highly productive agricultural areas. This is why most of the sugar, coffee, cotton, and other large-scale plantings of Central and South America, Indonesia, and the Philippines are situated downwind from recently active volcanoes.

From time to time, we have an opportunity to witness the manner in which plant life develops in volcanic regions. One of the best recent examples began on November 14, 1963, when the new island of Surtsey rose from the sea near the southern coast of Iceland and soon attained a surface area of about 3 km². Before long, the scoria and barren lava were colonized by species of plants and animals that were well adapted to the harsh ecological conditions, but with time, weathering, the decay of plants, and accumulation of windblown dust combined to produce a more extensive soil layer that supports increasingly diverse forms of life.

The rate of soil development on fresh volcanic rocks varies widely with climate and the texture and composition of the material. Lavas weather more slowly than ash, but well-drained, coarse scoria or pumice weathers more slowly than fine ash in areas of poor drainage. On the leeward slopes of the Hawaiian Islands, weathering may take 10 to 20 times longer than on the wetter, windward slopes. Even under the most favorable climatic conditions, however, it would take thousands of years for soils to develop on lava flows to depths great enough to support crops, while only a few decades are needed for fine powdery ash.

Other things being equal, the rate of soil development varies inversely with the silica content of the original material. Basaltic ash erupted from some of the Central American volcanoes is often planted with corn the year after it falls. Ash of intermediate composition, such as andesite, takes 10 or 20 years to develop soils under humid conditions. Healthy forests were growing on the andesitic ash deposits of

Fig. 14.8 The growth rings of this pine tree show the effect of the ash fall from Mt. Katami in 1912 on the rate of growth of vegetation. Diameter is 3 cm. (Photograph by R. F. Griggs.)

the Soufrière volcano of the West Indian island of St. Vincent about 30 years after the eruption of 1902, but the siliceous rhyolitic pumice that covers much of Guatemala and El Salvador shows only slight alteration after 2,000 years and requires up to 5,000 years to develop a weathered zone deep enough to support crops.

The ideal condition is that in which a light blanket of ash is laid down every decade or two. Where this ash is rich in potassium, phosphorus, and other essential elements, the rejuvenating effect may be startling (Fig. 14.8). Even in regions that do not receive this natural fertilizer because they are upwind from the volcanoes, it is becoming economically feasible to add ash to the clay-rich soils artifically and thereby increase the value of otherwise marginal farm land.

We should not close without mentioning the less tangible esthetic assets of volcanism and the magnificent, varied landscape it has given us. It is difficult to place a monetary value on the experience thousands of us enjoy when we visit the National Parks of Crater Lake, Mt. Rainier, or Yellowstone. Since the dawn of history, humans have attached special importance to these natural wonders that continue to remind us of the power and beauty of nature.

Suggested Reading

Blong, R. J. 1984. *Volcanic hazards*. Orlando: Academic Press.
A thorough treatment of problems of risk evalution and mitigation.

De Vivo, B., R. Scandone, and R. Trigila. 1993. Mount Vesuvius. *J Volc Geoth Res* (Spec Issue) 58: 381 p.
A comprehensive account of one of the most thoroughly studied volcanoes of the world and the risks of future eruptions.

Fiske, R. S. 1984. Volcanologists, journalists, and the concerned local public: A tale of two crises in the eastern Caribbean. In: *Explosive volcanism: Inception, evolution, and hazards*. Washington, DC: National Academic Press, pp. 170–6.
A thoughtful study of the role of volcanologists in dealing with public officials and the press during volcanic events that entail substantial risk.

Peterson, D. W. 1988. Volcanic hazards and public response. *J Geoph Res* 93:4161–70.
An analysis of the problems of mitigating the effects of volcanic eruptions.

Scandone, R., G. Arganese, and F. Galdi. 1993. The evaluation of volcanic risk in the Vesuvius area. *J Volc Geoth Res* (Spec Issue) 58:26–73.
An example of how volcanic risk is evaluated for a major metropolitan area.

Scarpa, R., and R. I. Tilling, eds. 1997. *Monitoring amd mitigation of volcanic hazards*. Berlin: Springer-Verlag.
A collection of papers dealing with techniques of assessing risk and mitigating the effects of volcanic eruptions.

Sheets, P. D., and D. K. Grayson, eds. 1979. *Volcanic activity and human ecology*. Academic Press, 396 p.
A collection of outstanding papers dealing with all aspect of the effects of volcanism on humans.

Tilling, R. I., and L. Wilson. 1989. *Volcanic hazards, Short course in geology. Am Geoph Union* 1: 123 p.
A review of the techniques of monitoring volcanoes and mitigating the effects of various types of hazards.

G L O S S A R Y

▲

Aa: the Hawaiian name for rough, scoriaceous lava.

Aerosol: very fine droplets or particles of volatile material that exsolve during strong explosive eruptions and condense on cooling.

Andesite: a calc-alkaline volcanic rock with 52 to 63% SiO_2 consisting mainly of plagioclase and pyroxene. Approximately midway in composition between basalt and rhyolite. Characteristic of subduction-related volcanoes.

Ash: fine explosive ejecta less than 2 mm in maximum dimension.

Asthenosphere: the layer of the Earth immediately below the lithosphere, which is relatively weak and readily deformed. Contains a small proportion of silicate melt.

Basalt: very common type of volcanic rock with a silica content of about 44 to 52%.

Basanite: basalt with a modal feldspathoid, normally nepheline.

Block: an erupted fragment of solid rock measuring more than 64 mm in maximum dimension.

Bomb: a mass of magma, normally rounded in form, that measures more than 64 mm in maximum dimension and is largely plastic when erupted.

Breccia: a mass of angular fragments of volcanic rock, normally with a wide range of sizes.

Calc-alkaline: a type of subalkaline magma series characterized in the middle stages of differentiation by andesites.

Caldera: a large volcanic depression, more or less circular or cirque-like in form, the diameter of which is many times greater than that of the included vent or vents. Small depressions formed by collapse are called pit-craters.

Carbonatite: a magma or igneous rock consisting mainly of carbonates. The most common minerals are calcite, magnesite, and sodium carbonate.

Chondrite: a common type of stony meteorite characterized by chondrules in a fine-grained matrix of silicate minerals and metallic nickel–iron. Their compositions differ somewhat, but on average they are probably close to the original undifferentiated compositions of the terrestrial planets.

Chondrule: a spherical grain or aggregate, often with radially oriented crystals of olivine and pyroxene found in many stony meteorites. Chondrules are thought to have formed from silicate droplets during accretion of the parent body of meteorites.

Cinders: *see* Scoria.

Coulée: a thick flow of lava with a steep front and margins.

Crater: a more or less circular depression up to 1 or 2 km in diameter at the summit or on the flanks of a volcano. May result either from explosions or collapse (*see* Caldera).

Crystal fractionation: the mechanical process of separation of crystals from a host magma.

Dacite: a felsic subalkaline volcanic rock with about 63 to 68% SiO^2 and at least 10% normative or modal quartz.

Diabase: *see* Dolerite.

Differentiation: the process of producing a magma or series of magmas with compositions that differ from that of the original magma from which they are derived.

Dike: A near-vertical, sheet-like magmatic intrusion.

Diorite: an intermediate plutonic rock in which neither quartz nor alkali feldspar account for more than 10% of the volume.

Dolerite: a medium-grained hypabyssal rock of basaltic composition; commonly found in dikes and sills. In America, synonymous with diabase.

Dome: a steep-sided, more or less round extrusion or shallow intrusion of viscous lava.

Dunite: a coarse-grained ultramafic rock composed primarily of olivine.

Eclogite: a high-pressure assemblage of sodic pyroxene and pyrope-rich garnet. The bulk chemical composition is similar to that of basalt.

Ferrobasalt: a basalt, normally tholeiitic, that has evolved to a high degree of iron enrichment with more than 12% total iron oxides.

Ferrogabbro: the plutonic equivalent of ferrobasalt.

Flood basalt: a voluminous, laterally extensive lava flow, normally erupted from a fissure.

Fumarole: a prolonged, nonexplosive release of steam and gases, usually rich in sulfur.

Gabbro: a coarse-grained rock of basaltic composition composed of calcic plagioclase, pyroxene, and possibly olivine, opaque oxides, and hornblende.

Geyser: a hot spring that intermittently erupts jets of hot water and steam.

Granite: a felsic plutonic rock consisting of about equal parts quartz, potassium feldspar, and sodic plagioclase. The name is also used as a general term for all leucocratic silica-rich plutonic rocks.

Granodiorite: a felsic plutonic rock similar to granite but with less potassium feldspar (less than a third of the total feldspar) and a higher proportion of dark minerals.

Guyot: a submarine volcano with a flat top produced by wave erosion before the island was submerged.

Harzburgite: a coarse-grained ultramafic rock composed primarily of olivine and orthopyroxene.

Hawaiite: an andesine-basalt. May have a small amount of normative nepheline.

Hotspot: a persistent source of magma at a fixed location in the mantle.

Hyaloclastite: glassy volcanic fragments resulting from sudden quenching in water.

Hydrothermal: an adjective denoting water that has been heated by a magmatic source at depth.

Icelandite: a tholeiitic volcanic rock similar to andesite but with more iron and less alumina (usually $<16.5\%$ Al_2O_3).

Ignimbrite: a fragmental volcanic rock laid down by a pyroclastic flow; most are siliceous and many are welded, but these are not essential elements of the definition.

Komatiite: an ultramafic lava with more than 20% MgO and consisting almost entirely of olivine and pyroxene.

Lahar: a type of fragmental volcanic material deposited by debris flows; characterized by unsorted clasts with a wide range of sizes. In older literature, lahars were referred to as "mudflows."

Lapilli: fragmental volcanic ejecta between 2 and 64 mm in diameter.

Lherzolite: a peridotite consisting of olivine, diopsidic augite, orthopyroxene, and possibly spinel or garnet.

Lithosphere: the uppermost 100 to 200 km of the solid Earth. Includes the crust and part of the upper mantle.

Maar: a low-rimmed explosion crater formed by a single phreatic or phreatomagmatic eruption. Commonly contains a lake.

Magma: a completely or partly molten natural substance that, on cooling, solidifies as a crystalline or glassy igneous rock. Contains dissolved volatile components.

Monzonite: a plutonic rock of intermediate silica content and sub-equal alkali feldspar and plagioclase; has less than 10% quartz.

Mudflow: *see* Lahar.

Norite: a type of gabbro containing hypersthene and plagioclase; augite and iron–oxides are also possible minerals.

Nuée ardente: a swiftly flowing cloud consisting of hot pyroclastic debris suspended in turbulent gas.

Obsidian: a siliceous volcanic rock consisting of glass with few if any crystals.

Oceanic ridge: a long system of dikes and fissures from which lava is erupted. An axis of spreading between two diverging oceanic plates.

Ophiolite: a collective name applied to basalts, gabbros, ultramafic rocks, and sediments that were originally part of the oceanic lithosphere and upper mantle.

Pahoehoe: a Hawaiian name for lava with a smooth, commonly corded surface.

Palagonite: hydrated basaltic glass, usually with a golden orange color, formed by subaqueous or subglacial eruptions.

Peridotite: an ultramafic plutonic rock consisting of at least 90% combined olivine and pyroxene.

Phenocryst: relatively large crystals in a somewhat finer-grained igneous rock.

Phonolite: an alkaline volcanic rock composed of alkali feldspar (typically anorthoclase or sanidine) and nepheline; the volcanic or hypabyssal equivalent of nepheline syenite.

Phreatic: an adjective denoting shallow meteoric water. In volcanic environments, may be superheated and capable of producing explosive eruptions.

Pillow lava: bulbous or tubular lava formed by subaqueous eruptions.

Porphyritic: an igneous rock with phenocrysts.

Pumice: highly vesicular siliceous glass.

Pyroclastic: an adjective denoting fragmentation by explosive volcanism.

Rhyolite: a felsic volcanic rock with more than 68% SiO^2; broadly equivalent to granite.

Scoria: highly vesicular pyroclastic ejecta, most commonly of basaltic or andesitic composition.

Sideromelane: clear basaltic glass.

Stratocone: a large steep-sided volcano constructed of relatively viscous lavas and pyroclastic debris.

Subduction: a process in which an oceanic lithospheric plate, normally of oceanic character, descends beneath another plate that may be either oceanic or continental.

Sublimations: precipitation of sulfur-rich deposits from gas.

Syenite: a felsic plutonic rock with subequal amounts of sodic plagioclase and K-feldspar but less than 10% quartz.

Tephra: a collective term for all types of pyroclastic deposits.

Tholeiite: a type of subalkaline magma characterized by normative hypersthene and strong iron enrichment in the middle stages of differentiation. The same name is applied to the basaltic parent of such a series.

Trachyte: a felsic volcanic rock with no nepheline and less than 10% quartz or tridymite; the volcanic equivalent of syenite.

Tsunami: a sea wave generated by a sudden displacement of the water by an event, such as an earthquake or volcanic avalanche.

Tuff: consolidated volcanic ash and lapilli.

Vesicle: bubbles of air or gas in a volcanic rock.

Volcanic arc: a chain of volcanic island associated with an oceanic trench and subduction zone.

Xenocryst: a crystal of foreign origin included in a magma or igneous rock.

Xenolith: a rock of foreign origin included in a magma or igneous rock.

REFERENCES

▲

Allen, E. T., and A. L. Day. 1935. Hot springs of the Yellowstone National Park. *Carnegie Inst Wash Pub* 466: 525 p.

Angell, J. K. 1990. Variations in global tropospheric temperature after adjustment for the El Nino influence, 1958–89. *Geophys Res Ltrs* 17(8):1093–6.

Bacon, C. R. 1982. Time-predictable bimodal volcanism in the Coso Range, California. *Geology* 10:65–9.

Bailey, R. A., G. E. Dalrymple, and M. A. Lanphere. 1976. Volcanism, structure and geochronology of Long Valley Caldera, Mono County, California. *J Geoph Res* 81:725–44.

Bailey, R. A., and D. P. Hill. 1990. Magmatic unrest at Long Valley Caldera, California, 1980–1990. *Geoscience Canada* 17:175–9.

Bardintzeff, J. M. 1985a. Les nuées ardentes: Pétrogenèse et volcanologie. Thèse de doctorat d'Etat, université de Paris-Sud, Orsay. *Bull PIRPSEV— CNRS* 109b.

Bardintzeff, J. M. 1985b. Calc-alkaline nuées ardentes: A new classification. *J Geodynamics* 3:303–25.

Bardintzeff, J. M. 1986. *Volcans et magmas*. Paris and Monaco: Le Rocher, 160 p.

Bardintzeff, J. M. 1993. *Volcans*. Paris: Armand Colin, 184 p.

Bardintzeff, J. M. 1997. *Connaitre et découvrir les volcans*. Geneva and Paris: Liber, 216 p.

Bardintzeff, J. M., J. Demange, and A. Gachon. 1986. Petrology of the volcanic bedrock of Mururoa atoll (Tuamotu archipelago, French Polynesia). *J Volc Geoth Res* 28:55–83.

Bardintzeff, J. M., and B. Bonin. 1987. The amphibole effect: A possible mechanism for triggering explosive eruptions. *J Volc Geoth Res* 33:255–62.

Bardintzeff, J. M., H. Bellon, B. Bonin, R. Brousse, and A. R. McBirney. 1988. Plutonic rocks from Tahiti Nui caldera (Society Archipelago, French Polynesia): A petrological, geochemical, and mineralogical study. *J Volc Geoth Res* 35:31–53.

Bardintzeff, J. M., J. C. Miskovsky, H. Traineau, and D. Westercamp. 1989. The recent pumice eruptions of Mt. Pelée Volcano, Martinique. Part II: Grain-size studies and modeling of the last Plinian phase P1. *J Volc Geoth Res* 38:35–48.

Barth, T. F. W. 1950. Volcanic geology, hot springs, and geysers of Iceland. *Carnegie Instit Wash Pub* 587: 174 p.

Basilevsky, A. T. 1990. Volcanism and tectonics on the Solar System planets and satellites: Dependence on the body radius and distance to the central body. *Twelfth Soviet-American Microsymposium*, Moscow, 16–17.

Bunsen, R. 1851. Ueber die Prozesse der Vulkanischen Gesteinbildungen Islands. *Pogg Ann* 83:197–272.

Carey, S., and H. Sigurdsson. 1987. Temporal variations in column height and magma discharge rate during the 79 AD eruption of Vesuvius. *Geol Soc Am Bull* 99:303–14.

Caron, J.-M., A. Gautier, A. Schaff, J. Ulysse, and J. Wozniak. 1995. *Comprendre et enseigner la planète Terre*, 3rd ed. Paris: Ophrys.

Cepeda, H., R. James, L. A. Murcia, E. Parra, R. Y. Salinas and H. Vergara. 1985. Mapa de riesgos volcanicos potenciales del Nevado del Ruiz, esc. 1/100,000. *Memoria explicativa* 1–28.

Chaigneau, M., H. Tazieff, and R. Fabre. 1960. Composition des gaz volcaniques du lac de lave permanent du Nyiragongo (Congo belge). *Comptes Rendu, Acad Sci Paris* 250:2482–5.

Chorowicz, J. 1983. Le rift E-africain: Début d'ouverture d'un océan. *Bull Elf Aquit* 7:155–62.

Cigolini, C., A. Borgia, and L. Casertano. 1984. Intra-crater activity, aa-block lava, viscosity and flow dynamics: Arenal Volcano, Costa Rica. *J Volc Geoth Res* 20:155–76.

Cox, K. G. 1978. Kimberlite pipes. *Sci Am* 238:4, 120–32

Duffield, W. A., J. H. Sass, and M. L. Sorey. 1994. Tapping the Earth's natural heat. *U S Geol Surv Circ* 1125: 63 p.

Duncan, R. A., J. Backman, and L. Peterson. 1989. *J Volc Geoth Res* 36:193–8.

Fenner, C. N. 1920. The Katmai region, Alaska, and the great eruption of 1912. *J Geol* 28:569–606.

Fink, J. H., S. O. Park, and R. Greeley. 1983. Cooling and deformation of sulfur flows. *Icarus* 56:38–50.

Fisher, R. V. 1964. Maximum size, median diameter, and sorting of tephra. *J Geoph Res* 69:341–55.

Fisher, R. V., and H. U. Schminke. 1984. *Pyroclastic rocks*. Berlin: Springer-Verlag, 472 p.

Fiske, R. S., C. A. Hopson, and A. C. Waters. 1963. Geology of Mount Rainier National Park. *U S Geol Surv Prof Paper* 444: 93 p.

Hill, D. P., R. A. Bailey, and A. S. Ryall. 1985. Active tectonic and magmatic processes beneath Long Valley caldera, Eastern California: An overview. *J Gephys Res* 90:B13, 11111–20.

Ishihara, K., M. Iguchi, and K. Kamo. 1990. Numerical simulations of lava flows on some volcanoes in Japan. In J. Fink, ed., *Lava flows and domes*. Berlin: Springer-Verlag, pp. 174–207.

Jaggar, T. A. 1947. Origin and development of craters. *Geol Soc Am Mem* 21: 508 p.

Kingsley, L. 1931. Cauldron subsidence of the Ossipee Mountains. *Am J Sci* 22:139–68.

Kushiro, I. 1973. Origin of some magmas in oceanic and circum-oceanic regions. *Tectonophys* 17:211–22.

Kushiro, I., and H. Kuno. 1963. Origin of primary basalt magmas and classification of basaltic rocks. *J Petrol* 4:75–89.

Lacroix, A. 1904. *La Montagne Pelée et ses éruptions*. Paris: Masson, 662 p.

Lipman, P. W. 1984. The roots of ash flow caldera in western North America: Windows into the tops of granitic batholiths. *J Geoph Res* 89:8801–41.

Lorenz, V. 1986. On the growth of maars and diatremes and its relevance to the formation of tuff rings. *Bull Volc* 48:265–74.

Macdonald, G. A. 1972. *Volcanoes*. Englewood Cliffs, NJ: Prentice Hall, 510 p.

Matsuo, S. 1961. The behavior of volatiles in magma. *J Earth Sci Nagoya Univ* 9:101–13.

Moore, J. G., and W. C. Albee. 1981. The 1980 eruptions of Mount St. Helens, Washington: Topographic and structural changes. In P. W. Lipman and D. R. Mullineaux, eds., The 1980 eruptions of Mount St. Helens, Washington. *U S Geol Surv Prof Paper* 1250:123–34.

Moore, J. G., R. L. Phillips, R. W. Grigg, D. W. Peterson, and D. A. Swanson. 1973. Flow of lava into the sea, 1969–1971, Kilauea volcano, Hawaii. *Geol Soc Am Bull* 84:537–46.

Morin, S., and J. Pahai. 1987. La catastrophe de Nyos (Cameroun). *Rev Géogr Cameroun* 6:81.

Mullineaux, D. R., D. W. Peterson, and D. R. Crandell. 1987. Volcanic hazards in the Hawaiian Islands. *U S Geol Surv Prof Paper* 1350:599–621.

Mysen, B. O., and A. L. Boettcher. 1975. Melting of a hydrous mantle. I: Phase relations of natural peridotite at high pressures and temperatures with controlled activities of water, carbon dioxide, and hydrogen. *J Petrol* 16:520–48.

Nakada, S. 1993. Unzen. *Bull Global Volcanism Network*, Smithsonian Institution, 18,5:8–9.

Nakada, S., and T. Fujii. 1993. Preliminary report on the activity at Unzen Volcano (Japan), November 1990–November 1991: Dacite lava domes and pyroclastic flows. *J Volc Geoth Res* 54:319–33.

Newhall, C. G., and S. Self. 1982. The volcanic explosive index (VEI): An estimate of explosive magnitude for historical volcanism. *J Geoph Res* 87-C2:1231–8.

Ninkovich, D., R. S. J. Sparks, and M. T. Ledbetter. 1978. The exceptional magnitude and intensity of the Toba eruption, Sumatra: An example for the use of deep-sea tephra layers as a geological tool. *Bull Volc* 41-3:286–98.

Nordlie, B. E. 1971. The composition of the magmatic gas of Kilauea and its behavior in the near-surface environment. *Am J Sci* 271:417–63.

Palumbo, A. 1998. Long-term forecasting of the extreme eruptions of Etna. *J Volc Geoth Res* 83:167–71.

Peck, D. 1978. Cooling and vesiculation of Alae lava lake, Hawaii. *U S Geol Surv Prof Paper* 935B.

Perret, F. A. 1924. The Vesuvius eruption of 1906. *Carnegie Inst Washington Pub* 339:151.

Pierson, T. C., R. J. Janda, J. C. Thouet, and C. A. Borrero. 1990. Perturbation and melting of snow and ice by the 13 November 1985 eruption of Nevado del Ruiz Volcano, Columbia, and consequent mobilization, flow and deposition of lahars. *J Volc Geoth Res* 41:17–66.

Pyle, D. M. 1990. New estimates for the volume of the Minoan eruption. *Thera and the Aegean World III* 2:113–21.

Ringwood, A. E. 1966. Phase transformation and differentiation in subducted lithosphere: Implications for mantle dynamics, basalt petrogenesis and crustal evolution. *J Geol* 90:611–43.

Romano, R., T. Caltabiano, P. Carveni, and M. F. Grasso. 1992. Etna. *Bull Global Volc Network*, Smithsonian Institution 17:4.

Sarna-Wojcicki, A., S. Shipley, R. B. Waitt, D. Dzurizin, and S. H. Wood. 1981. Areal distribution, thickness, mass, volume and grain-size of air-fall ash from six major eruptions of 1980. In P. W. Lipman and D. R. Mullineaux, eds., The 1980 eruptions of Mount St. Helens, Washington. *U S Geol Surv Prof Paper* 1250:577–600.

Savage, J. C., R. S. Cockerham, J. E. Estrem, and L. R. Moore. 1987. Deformation near Long Valley caldera, eastern California, 1882–1986. *J Geoph Res* 92:2721–46.

Scandone, R., G. Arganese, and F. Galdi. 1993. The evaluation of volcanic risk in the Vesuvian area. *J Volc Geoth Res* 58:261–73.

Shepherd, E. S. 1938. The analysis of gases obtained from volcanoes and from rocks. *J Geol* 33:289–370.

Sigurdsson, H. 1987. Lethal gas bursts from Cameroon crater lakes. *EOS* 68(23):570–3.

Sigvaldason, G. E., and G. Elisson. 1968. Collection and analysis of volcanic gases at Surtsey, Iceland. *Geoch Cosmoch Acta* 32:797–805.

Simkin, T., and L. Siebert. 1994. *Volcanoes of the world*, 2d ed. Missoula, MN: Smithsonian Institute, Geoscience Press, 349 p.

Stroudsburg, PA: Hutchinson Ross, 240 p.

Smith, R. B., and L. W. Braile. 1994. The Yellowstone Hotspot. *J Volc Geoth Res* 61:121–87.

Sparks, R. S. J. 1976. Grain size variations in ignimbrites and implications for the transport of pyroclastic flows. *Sedimentology* 23:147–88.

Sparks, R. S. J. 1986. The dimensions and dynamics of volcanic eruption columns. *Bull Volc* 48:3–15.

Sparks, R. S. J., P. Meyer, and H. Sigurdson. 1980. *Earth & Planet Sci Ltrs* 46:419–30.

Stowe, L. 1993. Atmospheric effects. *Bull Global Volc Network*, Smithsonian Institution, 18:9.

Tazieff, H. 1984. Mt. Niragongo: Renewed activity of the lava lake. *J Volc Geoth Res* 20:267–80.

Thorarinsson, S., and G. E. Sigvaldason. 1962. The eruption of Askja, 1961: A preliminary report. *Am J Sci* 260:641–51.

Thorarinsson, S., and G. E. Sigvaldason. 1972. The Hekla eruption of 1970. *Bull Volc* 36:269–88.

Varet, J., and J. Demange. 1980. Autoclastic submarine breccias in hole 410, leg 49, and other DSDP sites. *Initial reports of the DSDP* XLIX:749–60.

Vuagnat, M. 1975. Pillow lava flows: Isolated sacks or connected tubes? *Bull Volc* 39:581–9.

Wadge, G. 1981. The variations of magma discharge during basaltic eruptions. *J Volc Geoth Res* 11:139–68.

Wadge, G., G. P. L. Walker, and J. E. Guest. 1975. The output of Etna volcano. *Nature* 255:385–7.

Walker, G. P. L. 1971. Grain size characteristics of pyroclastic deposits. *J Geol* 79:696–714.

Walker, G. P. L. 1973. Explosive volcanic eruptions: A new classification scheme. *Geol Rundschau* 62:431–46.

Walker, G. P. L. 1981. Plinian eruptions and their products. *Bull Volc* 44: 223–40.

Westercamp, D. 1980. Une méthode d'évaluation et de zonation des risques volcaniques à la Soufrière de Guadeloupe, Antilles, Françaises. *Bull Volc* 43:431–52.

Westercamp, D., and H. Traineau. 1983. The past 5000 years of volcanic activity at Mount Pelée, Martinique (F. W. I.): Implications for assessment of volcanic hazards. *J Volc Geoth Res* 17:159–85.

White, D. E. 1968. Hydrology, activity, and heat flow of the Steamboat Springs thermal system, Washoe County, Nevada. *U S Geol Surv Prof Paper* 458-C: 109 p.

Williams, H. 1942. Geology of Crater Lake National Park, Oregon. *Carnegie Inst Wash Pub* 540: 162 p.

Wilson, L., R. S. J. Sparks, and G. P. L. Walker. 1980. Explosive volcanic eruptions. IV: The control of magma properties and conduit geometry on eruption column behavior. *Geophys J Roy Astron Soc* 63:117–48.

Wohletz, K. H., and M. F. Sheridan. 1983. Hydrovolcanic explosions. II: Evolution of basaltic tuff rings and tuff cones. *Am J Sci* 283:385–413.

Wright, J. V., A. L. Smith, and S. Self. 1980. A working terminology of pyroclastic deposits. *J Volc Geoth Res* 8:315–36.

INDEX

▲